STATISTICAL POCKETBOOK

WORLD FOOD AND AGRICULTURE

2022

Food and Agriculture Organization of the United Nations
Rome, 2022

CONTENTS

FOREWORD

Timely, accurate and high-quality data and statistics are the cornerstone of solid policy design, where decisions are based on hard evidence, and monitoring and evaluation rely on strong statistical systems. This has become all the more critical as governments around the world commit to major sectoral and national development plans, as well as regional and global development agendas. Furthermore, recent events like the COVID-19 pandemic and the war in Ukraine have emphasized the pressing need for data and statistics to inform timely responses and monitor trends.

Statistical work has been at the core of the activities and mandate of the Food and Agriculture Organization of the United Nations (FAO) since the Organization was founded in 1945, supporting its Members in eliminating hunger, improving nutrition, eradicating rural poverty, and promoting inclusive and efficient agrifood systems. FAO is a leading provider of internationally comparable data on food, nutrition and agriculture, which are gathered from national statistical offices and from FAO's network of partner agencies and are harmonized to paint a global picture.

This Pocketbook, prepared by the Statistics Division of FAO, provides quick and easy access to top-level numbers, charts and maps on the many dimensions of food and agriculture – ranging from the characteristics of the sector to production, prices and trade, as well as food security and nutrition, and environmental aspects. More than 50 indicators in 17 thematic domains for around 200 countries and regions are presented in this companion volume to the *World Food and Agriculture Statistical Yearbook 2022*.

In addition to compiling and disseminating data, FAO is also involved in strengthening the statistical capacity of countries in order to produce more and better data; setting standards and methodologies; and leveraging big data innovations. FAO is committed to ensuring free access to current, reliable, timely and trusted data, necessary to chart a course towards a more sustainable and equitable agrifood systems and a world free of hunger.

José Rosero Moncayo
Director, Statistics Division

HOW TO USE THIS POCKETBOOK

THE STRUCTURE

The Statistical Pocketbook 2022 presents selected key indicators related to agriculture and food security that the international community, governments, the private sector and civil society can use to assess current trends and prioritize their actions. It presents a variety of agriculture and food security dimensions along four main focus areas:

- An overview of agriculture, forestry and fishing from an economic standpoint, highlighting the use of the factors of production.
- The outputs of the sector in terms of production and trade of the different commodities and the evolution of prices.
- How some of these outputs are consumed by narrowing the focus on food security and nutrition.
- The impacts of the sector as a whole on the environment, in particular water and greenhouse gas emissions

This publication draws on the latest available data to describe through charts the trends since the early 2000s and show with maps the data for the latest year available.

COUNTRY DEFINITIONS AND CLASSIFICATION

The country classification adopted in this publication is based on the United Nations M49 classification *https://unstats.un.org/unsd/methodology/m49/*. The country names have been abbreviated in the data tables and figures. The official Food and Agriculture Organization of the United Nations (FAO) names can be found at *http://www.fao.org/nocs/en*.

AGGREGATIONS

Regional and subregional aggregates are based on the country groupings defined in the United Nations M49 classification. A small subset of indicators in the data tables is based on the aggregation rules defined in *The State of Food Security and Nutrition in the World 2022* report, which can be found at *https://doi.org/10.4060/cc0639en*.

Two types of aggregations are used: sum and weighted mean. A sufficiency condition is imposed when computing the aggregation – the aggregation is computed only when enough countries have reported data, and the current threshold is set at 50 percent of the variable and the weighting variable, if present.

DATA PRESENTATION CONVENTIONS

The cut-off date for the data is 6 October 2022.

- When country data have not been reported for the reference year, an asterisk (*) on the year label indicates that the value for the most recent year available is shown. For example, 2018–2020* means that the most recent value for the period from 2018 to 2020 is shown.
- A billion is 1 000 million.
- A trillion is 1 000 billion.

In data tables:

- A blank means that data are not available or that aggregates cannot be calculated because of missing data for the years shown.
- 0 or 0.0 means zero or a number that is small enough to round to zero at the displayed number of decimal places.
- <2.5 means a proportion less than 2.5 percent.
- <0.1 means less than 100 000 people.

1 VALUE ADDED

FIGURE 1. **VALUE ADDED OF AGRICULTURE, FORESTRY AND FISHING BY REGION**

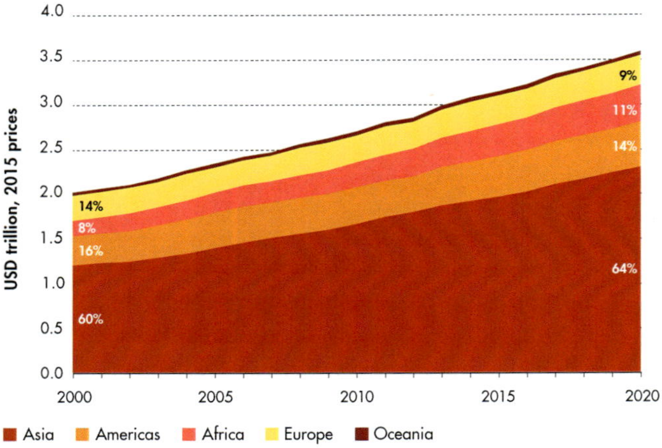

Note: Percentages on the figure indicate the shares in the total; they may not tally due to rounding.
Source: FAO. 2022. FAOSTAT: Macro Indicators. In: *FAO*. Rome. Cited October 2022.
http://www.fao.org/faostat/en/#data/MK
https://doi.org/10.4060/cc2211en-fig01

MAP 1. **VALUE ADDED OF AGRICULTURE, FORESTRY AND FISHING BY REGION (2020)**

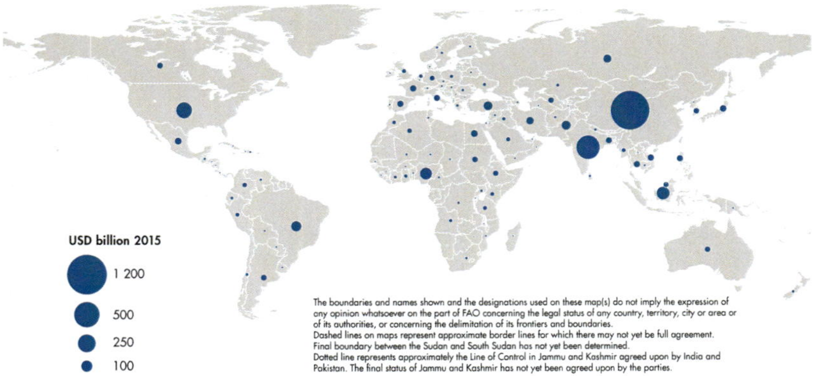

Source: FAO. 2022. FAOSTAT: Macro Indicators. In: *FAO*. Rome. Cited October 2022.
http://www.fao.org/faostat/en/#data/MK based on UN Geospatial. 2020. Map geodata [shapefiles]. New York, USA, UN.
https://doi.org/10.4060/cc2211en-map01

FIGURE 2. **SHARE OF AGRICULTURE, FORESTRY AND FISHING VALUE ADDED IN TOTAL GDP BY REGION (USD 2015 PRICES)**

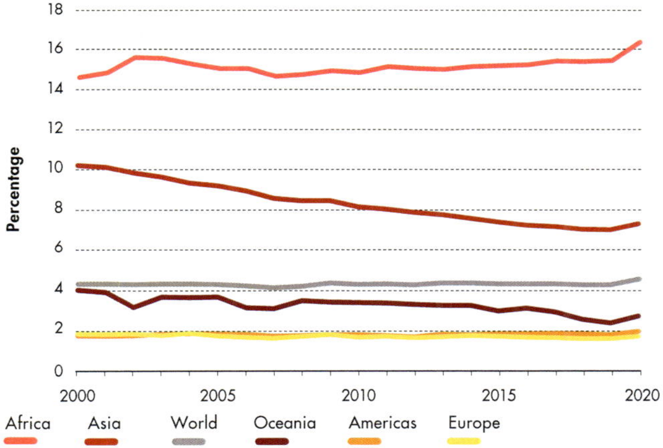

Source: FAO. 2022. FAOSTAT: Macro Indicators. In: *FAO*. Rome. Cited October 2022.
http://www.fao.org/faostat/en/#data/MK
https://doi.org/10.4060/cc2211en-fig02

MAP 2. **SHARE OF AGRICULTURE, FORESTRY AND FISHING VALUE ADDED IN TOTAL GDP (2020, USD 2015 PRICES)**

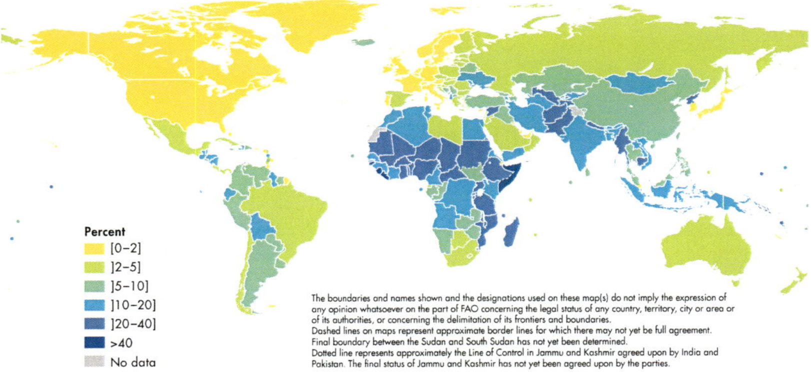

Source: FAO. 2022. FAOSTAT: Macro Indicators. In: *FAO*. Rome. Cited October 2022.
http://www.fao.org/faostat/en/#data/MK based on UN Geospatial. 2020. Map geodata [shapefiles]. New York, USA, UN.
https://doi.org/10.4060/cc2211en-map02

2 LAND USE

FIGURE 3. **SHARE OF LAND AREA BY TYPE AND REGION**

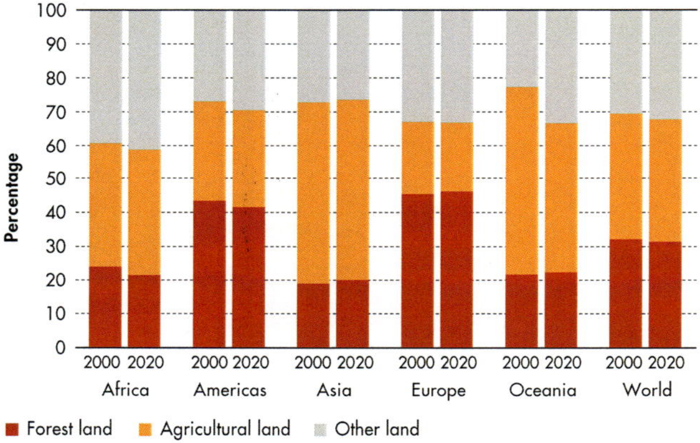

Source: FAO. 2022. FAOSTAT: Land Use. In: *FAO*. Rome. Cited October 2022.
http://www.fao.org/faostat/en/#data/RL
https://doi.org/10.4060/cc2211en-fig59

MAP 3. **CROPLAND AREA (2020)**

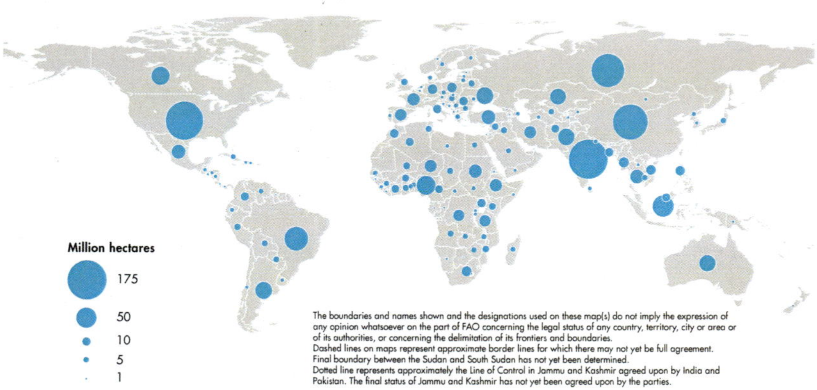

Source: FAO. 2022. FAOSTAT: Land Use. In: *FAO*. Rome. Cited October 2022.
http://www.fao.org/faostat/en/#data/RL based on UN Geospatial. 2020. Map geodata [shapefiles]. New York, USA, UN.
https://doi.org/10.4060/cc2211en-map05

FIGURE 4. **SHARE OF AGRICULTURAL LAND BY TYPE AND REGION**

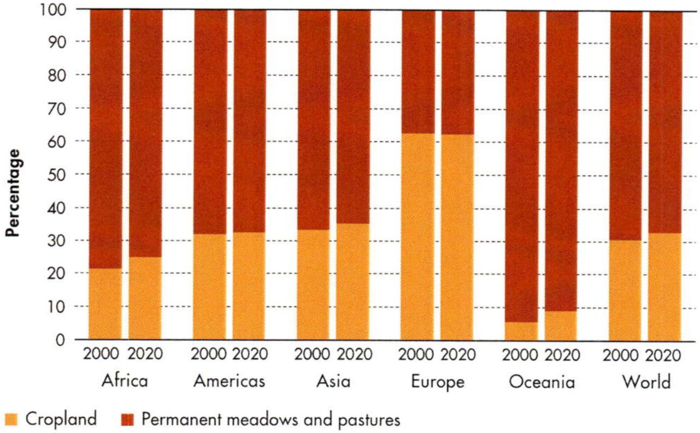

Source: FAO. 2022. FAOSTAT: Land Use. In: *FAO*. Rome. Cited October 2022.
http://www.fao.org/faostat/en/#data/RL
https://doi.org/10.4060/cc2212en-fig04

MAP 4. **CROPLAND AREA PER CAPITA (2020)**

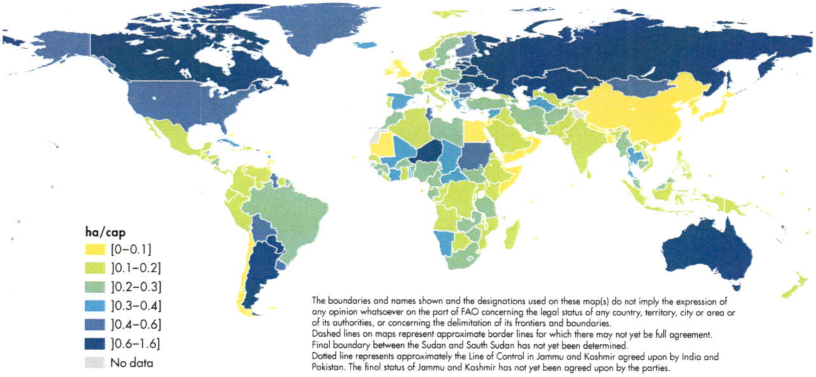

Source: FAO. 2022. FAOSTAT: Land use indicators. In: *FAO*. Rome. Cited October 2022.
http://www.fao.org/faostat/en/#data/EL based on UN Geospatial. 2020. Map geodata [shapefiles]. New York, USA, UN.https://doi.org/10.4060/cc2212en-map04

3 LABOUR

FIGURE 5. **EMPLOYMENT IN AGRICULTURE, FORESTRY AND FISHING BY REGION**

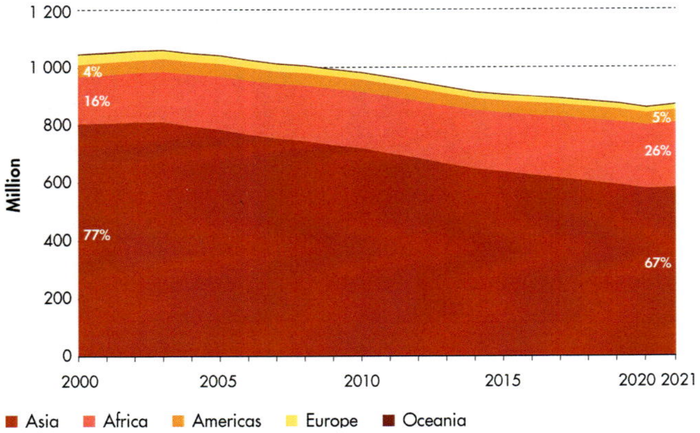

Note: Percentages on the figure indicate the shares in the total; they may not tally due to rounding.
Source: FAO. 2022. FAOSTAT: Employment Indicators: Agriculture. In: *FAO*. Rome. Cited October 2022.
http://www.fao.org/faostat/en/#data/OEA
https://doi.org/10.4060/cc2211en-fig10

MAP 5. **EMPLOYMENT IN AGRICULTURE, FORESTRY AND FISHING (2021)**

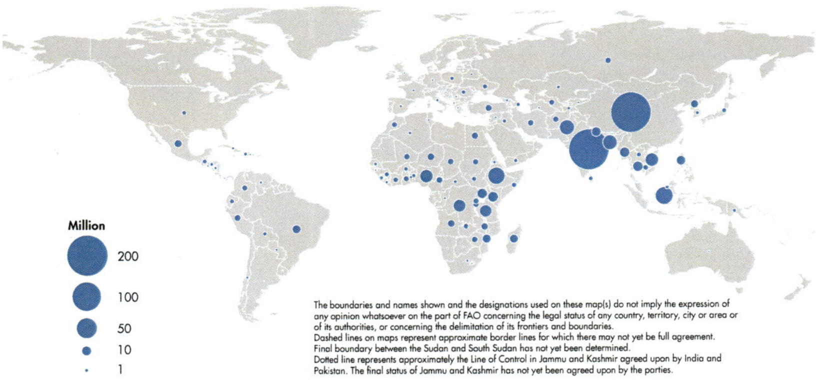

Source: FAO. 2022. FAOSTAT: Employment Indicators: Agriculture. In: *FAO*. Rome. Cited October 2022.
http://www.fao.org/faostat/en/#data/OEA based on UN Geospatial. 2020. Map geodata [shapefiles]. New York, USA, UN.
https://doi.org/10.4060/cc2211en-map03

FIGURE 6. **SHARE OF AGRICULTURE, FORESTRY AND FISHING EMPLOYMENT IN TOTAL EMPLOYMENT BY REGION**

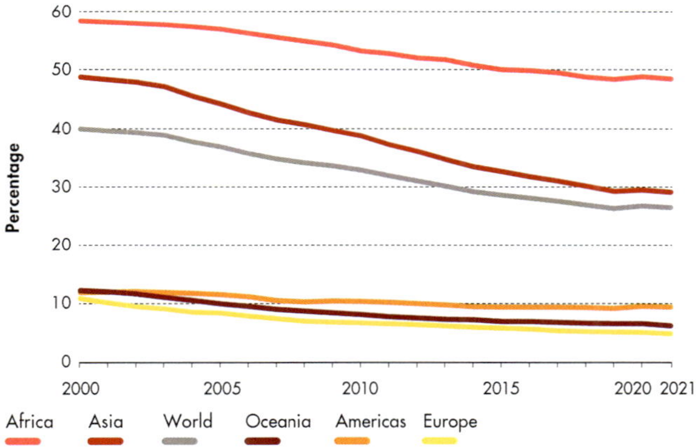

Source: FAO. 2022. FAOSTAT: Employment Indicators: Agriculture. In: *FAO*. Rome. Cited October 2022. http://www.fao.org/faostat/en/#data/OEA
https://doi.org/10.4060/cc2211en-fig11

MAP 6. **SHARE OF AGRICULTURE, FORESTRY AND FISHING EMPLOYMENT IN TOTAL EMPLOYMENT (2021)**

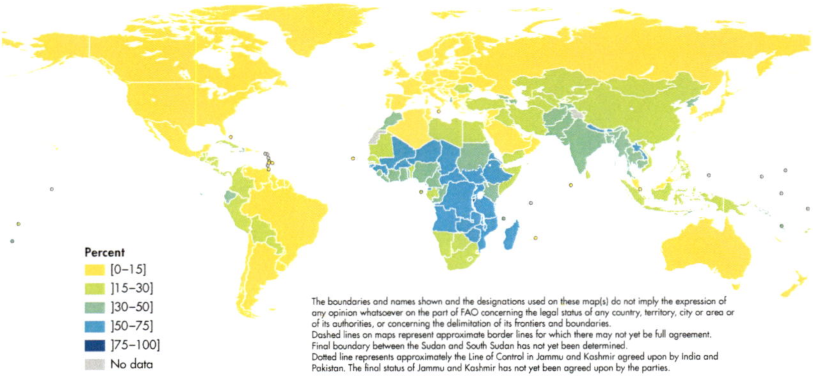

Source: FAO. 2022. FAOSTAT: Employment Indicators: Agriculture. In: *FAO*. Rome. Cited October 2022. http://www.fao.org/faostat/en/#data/OEA based on UN Geospatial. 2020. Map geodata [shapefiles]. New York, USA, UN.
https://doi.org/10.4060/cc2211en-map04

4 INVESTMENTS

FIGURE 7. **AID DISBURSEMENT FLOWS TO AGRICULTURE, FORESTRY AND FISHING BY RECIPIENT**

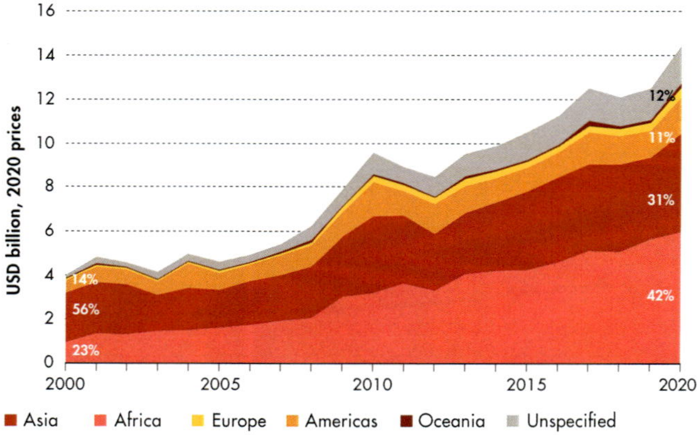

Note: Percentages on the figure indicate the shares in the total; they may not tally due to rounding.
Source: FAO. 2022. FAOSTAT: Development Flows to Agriculture. In: *FAO*. Rome. Cited October 2022.
http://www.fao.org/faostat/en/#data/EA
https://doi.org/10.4060/cc2212en-fig07

MAP 7. **RECIPIENTS OF AID DISBURSEMENT FLOWS TO AGRICULTURE, FORESTRY AND FISHING (2020)**

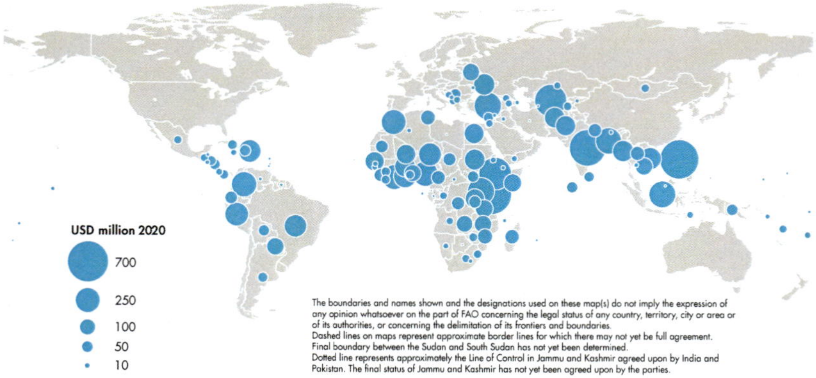

The boundaries and names shown and the designations used on these map(s) do not imply the expression of any opinion whatsoever on the part of FAO concerning the legal status of any country, territory, city or area or of its authorities, or concerning the delimitation of its frontiers and boundaries.
Dashed lines on maps represent approximate border lines for which there may not yet be full agreement.
Final boundary between the Sudan and South Sudan has not yet been determined.
Dotted line represents approximately the Line of Control in Jammu and Kashmir agreed upon by India and Pakistan. The final status of Jammu and Kashmir has not yet been agreed upon by the parties.

Source: FAO. 2022. FAOSTAT: Development Flows to Agriculture. In: *FAO*. Rome. Cited October 2022.
http://www.fao.org/faostat/en/#data/EA based on UN Geospatial. 2020. Map geodata [shapefiles]. New York, USA, UN.
https://doi.org/10.4060/cc2212en-map07

FIGURE 8. **GROSS FIXED CAPITAL FORMATION (AGRICULTURE, FORESTRY AND FISHING) AS A SHARE OF VALUE ADDED, TOP COUNTRIES BY REGION (2021, USD 2015 PRICES)**

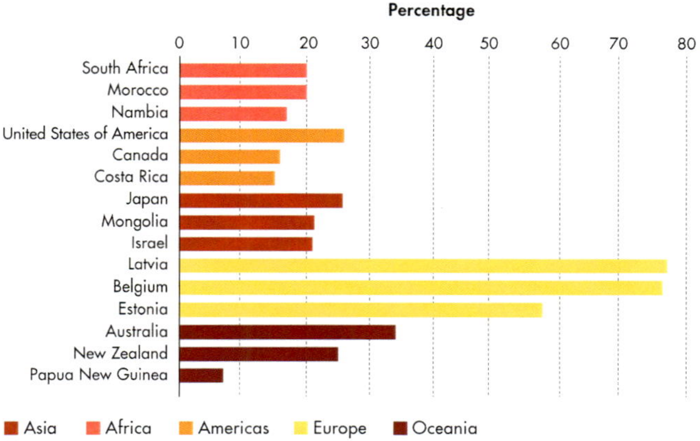

Note: Only countries with more than USD 500 million of agriculture value added are included.
Source: FAO. 2022. FAOSTAT: Capital Stock. In: *FAO*. Rome. Cited October 2022.
http://www.fao.org/faostat/en/#data/CS
https://doi.org/10.4060/cc2211en-fig13

MAP 8. **GROSS FIXED CAPITAL FORMATION (AGRICULTURE, FORESTRY AND FISHING) AS A SHARE OF VALUE ADDED (2021, USD 2015 PRICES)**

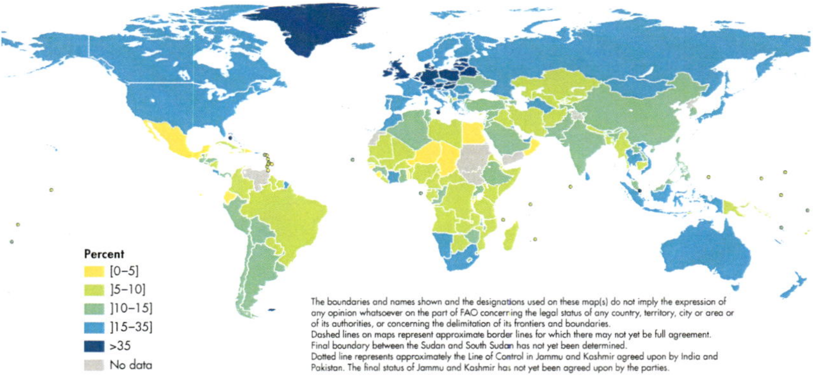

The boundaries and names shown and the designations used on these map(s) do not imply the expression of any opinion whatsoever on the part of FAO concerning the legal status of any country, territory, city or area or of its authorities, or concerning the delimitation of its frontiers and boundaries.
Dashed lines on maps represent approximate border lines for which there may not yet be full agreement.
Final boundary between the Sudan and South Sudan has not yet been determined.
Dotted line represents approximately the Line of Control in Jammu and Kashmir agreed upon by India and Pakistan. The final status of Jammu and Kashmir has not yet been agreed upon by the parties.

Source: FAO. 2022. FAOSTAT: Capital Stock. In: *FAO*. Rome. Cited October 2022.
http://www.fao.org/faostat/en/#data/CS based on UN Geospatial. 2020. Map geodata [shapefiles]. New York, USA, UN.
https://doi.org/10.4060/cc2211en-map06

5 PESTICIDES

FIGURE 9. **PESTICIDE USE BY REGION**

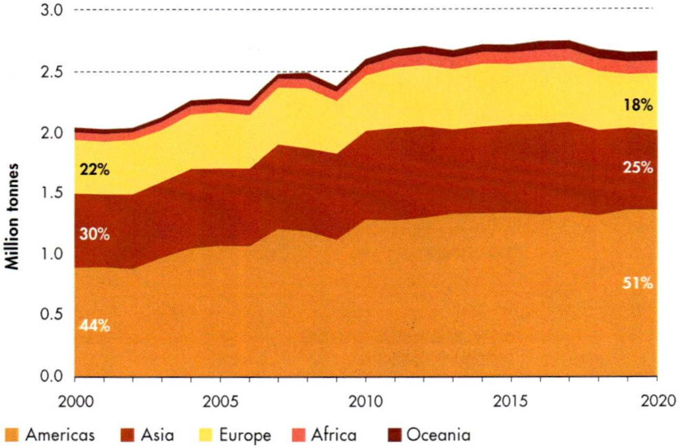

Note: Percentages on the figure indicate the shares in the total; they may not tally due to rounding.
Source: FAO. 2022. FAOSTAT: Pesticides Use. In: *FAO*. Rome. Cited October 2022.
http://www.fao.org/faostat/en/#data/RP
https://doi.org/10.4060/cc2211en-fig15

MAP 9. **PESTICIDE USE (2020)**

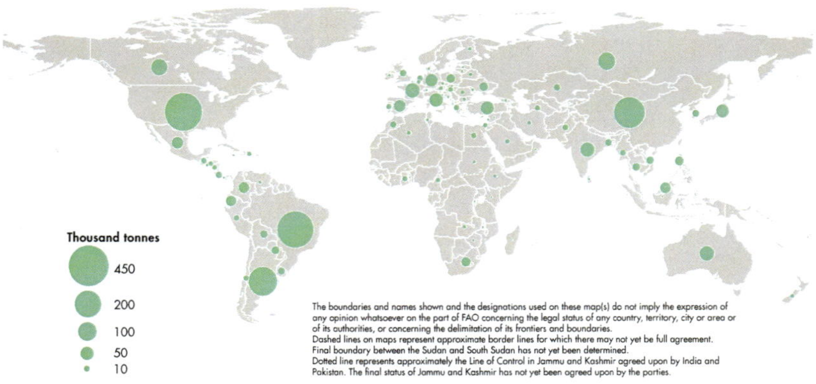

Source: FAO. 2022. FAOSTAT: Pesticides Use. In: *FAO*. Rome. Cited October 2022.
http://www.fao.org/faostat/en/#data/RP based on UN Geospatial. 2020. Map geodata [shapefiles]. New York, USA, UN.
https://doi.org/10.4060/cc2211en-map07

FIGURE 10. **PESTICIDE USE PER CROPLAND AREA BY REGION**

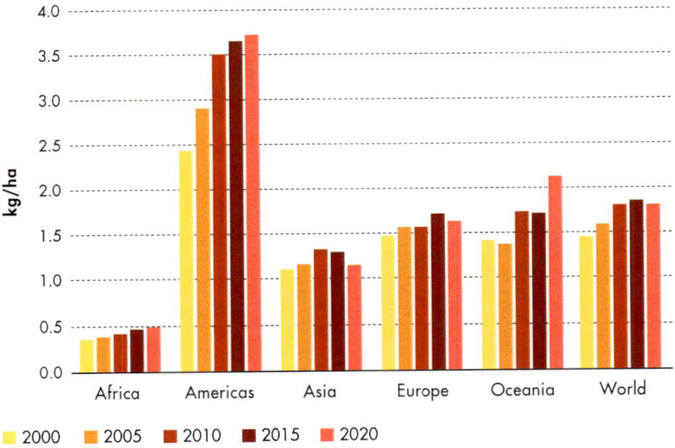

Source: FAO. 2022. FAOSTAT: Pesticides Indicators. In: *FAO*. Rome. Cited October 2022.
http://www.fao.org/faostat/en/#data/EP
https://doi.org/10.4060/cc2211en-fig16

MAP 10. **PESTICIDE USE PER CROPLAND AREA (2020)**

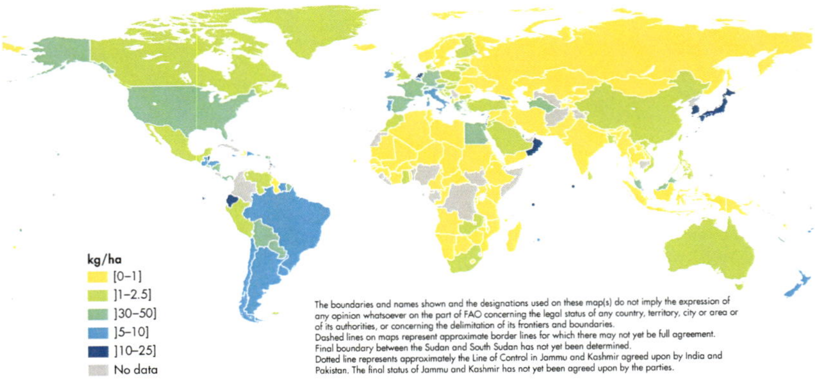

Source: FAO. 2022. FAOSTAT: Pesticides Indicators. In: *FAO*. Rome. Cited October 2022.
http://www.fao.org/faostat/en/#data/EP based on UN Geospatial. 2020. Map geodata [shapefiles]. New York, USA, UN.
https://doi.org/10.4060/cc2212en-map10

6 FERTILIZERS

FIGURE 11. **INORGANIC FERTILIZER USE BY REGION**

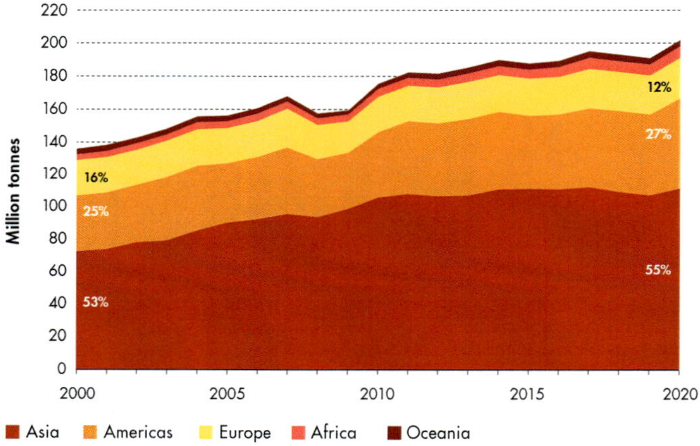

Note: Percentages on the figure indicate the shares in the total; they may not tally due to rounding.
Source: FAO. 2022. FAOSTAT: Fertilizers by Nutrient. In: *FAO*. Rome. Cited October 2022.
http://www.fao.org/faostat/en/#data/RFN
https://doi.org/10.4060/cc2211en-fig17

MAP 11. **INORGANIC FERTILIZER USE (2020)**

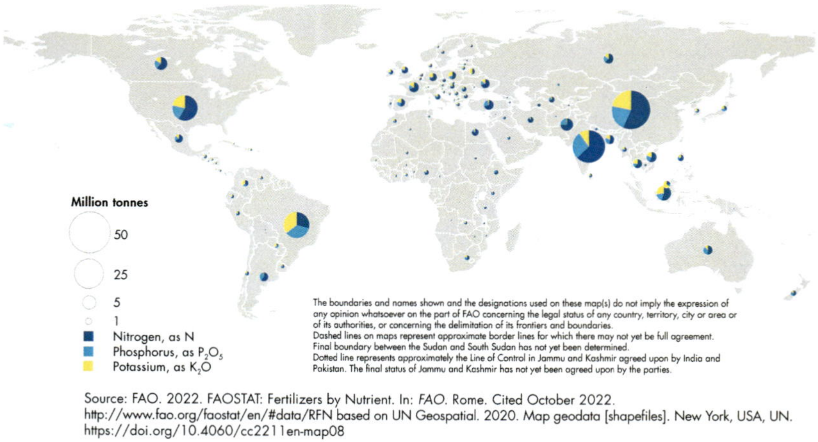

Source: FAO. 2022. FAOSTAT: Fertilizers by Nutrient. In: *FAO*. Rome. Cited October 2022.
http://www.fao.org/faostat/en/#data/RFN based on UN Geospatial. 2020. Map geodata [shapefiles]. New York, USA, UN.
https://doi.org/10.4060/cc2211en-map08

FIGURE 12. INORGANIC FERTILIZER USE PER CROPLAND AREA BY NUTRIENT AND REGION

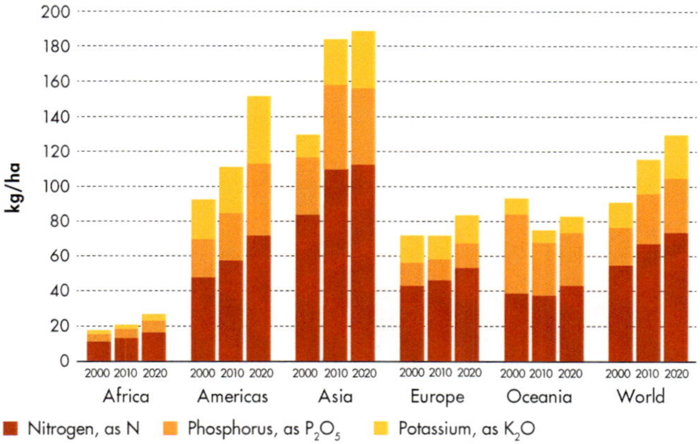

Source: FAO. 2022. FAOSTAT: Fertilizers by Nutrient. In: *FAO*. Rome. Cited October 2022.
http://www.fao.org/faostat/en/#data/RFN
https://doi.org/10.4060/cc2211en-fig19

MAP 12. INORGANIC FERTILIZER USE PER CROPLAND AREA (2020)

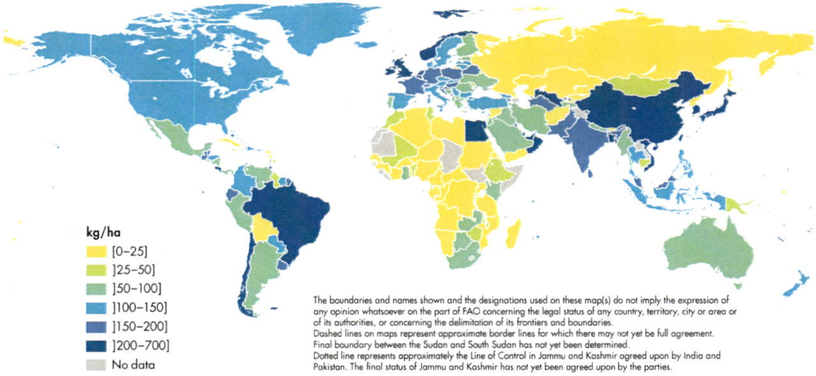

The boundaries and names shown and the designations used on these map(s) do not imply the expression of any opinion whatsoever on the part of FAO concerning the legal status of any country, territory, city or area or of its authorities, or concerning the delimitation of its frontiers and boundaries.
Dashed lines on maps represent approximate border lines for which there may not yet be full agreement.
Final boundary between the Sudan and South Sudan has not yet been determined.
Dotted line represents approximately the Line of Control in Jammu and Kashmir agreed upon by India and Pakistan. The final status of Jammu and Kashmir has not yet been agreed upon by the parties.

Source: FAO. 2022. FAOSTAT: Fertilizers by Nutrient. In: *FAO*. Rome. Cited October 2022.
http://www.fao.org/faostat/en/#data/RFN based on UN Geospatial. 2020. Map geodata [shapefiles]. New York, USA, UN.
https://doi.org/10.4060/cc2212en-map12

FIGURE 13. **WORLD PRODUCTION OF CROPS BY COMMODITY GROUP**

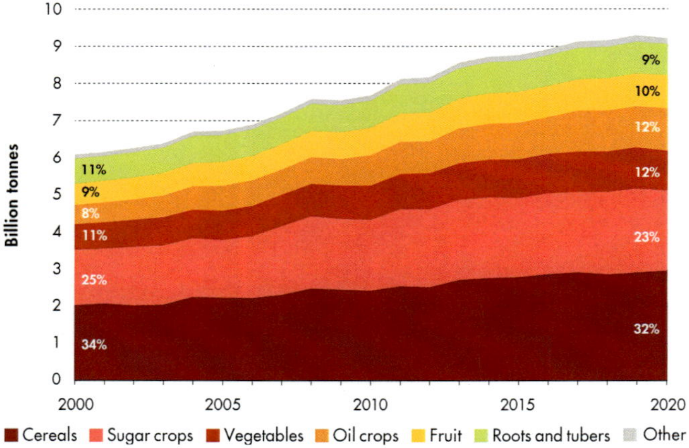

Note: Percentages on the figure indicate the shares in the total; they may not tally due to rounding.
Source: FAO. 2022. FAOSTAT: Production: Crops and livestock products. In: *FAO*. Rome. Cited October 2022.
https://www.fao.org/faostat/en/#data/QCL
https://doi.org/10.4060/cc2211en-fig20

MAP 13. **PRODUCTION OF CEREALS (2020)**

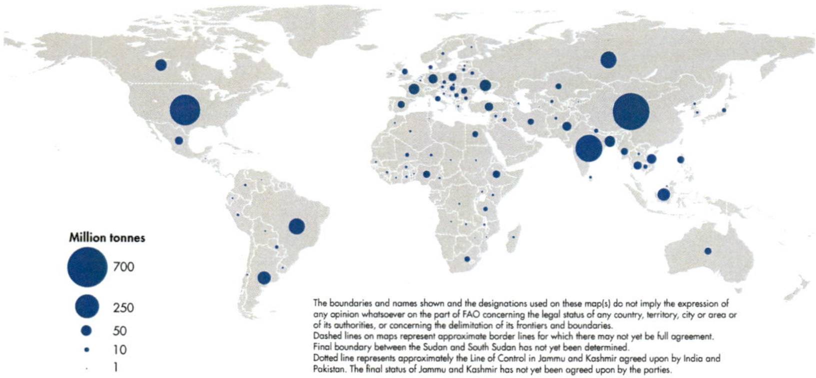

The boundaries and names shown and the designations used on these map(s) do not imply the expression of any opinion whatsoever on the part of FAO concerning the legal status of any country, territory, city or area or of its authorities, or concerning the delimitation of its frontiers and boundaries.
Dashed lines on maps represent approximate border lines for which there may not yet be full agreement.
Final boundary between the Sudan and South Sudan has not yet been determined.
Dotted line represents approximately the Line of Control in Jammu and Kashmir agreed upon by India and Pakistan. The final status of Jammu and Kashmir has not yet been agreed upon by the parties.

Source: FAO. 2022. FAOSTAT: Production: Crops and livestock products. In: *FAO*. Rome. Cited October 2022.
https://www.fao.org/faostat/en/#data/QCL based on UN Geospatial. 2020. Map geodata [shapefiles]. New York, USA, UN.
https://doi.org/10.4060/cc2211en-map09

FIGURE 14. **WORLD PRODUCTION OF CROPS, MAIN COMMODITIES**

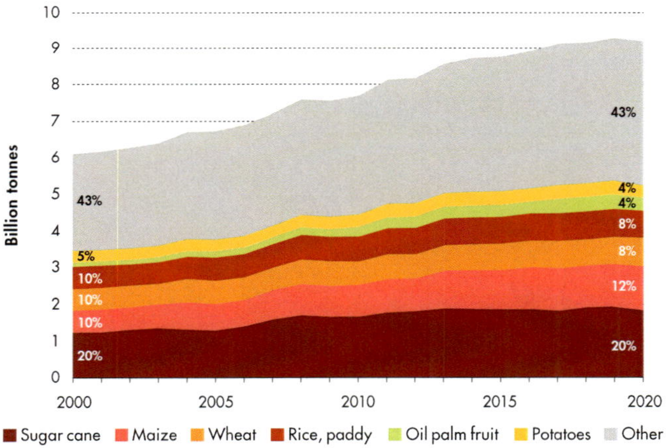

Note: Percentages on the figure indicate the shares in the total; they may not tally due to rounding.
Source: FAO. 2022. FAOSTAT: Production: Crops and livestock products. In: *FAO*. Rome. Cited October 2022.
https://www.fao.org/faostat/en/#data/QCL
https://doi.org/10.4060/cc2211en-fig21

MAP 14. **PRODUCTION OF SUGAR CANE (2020)**

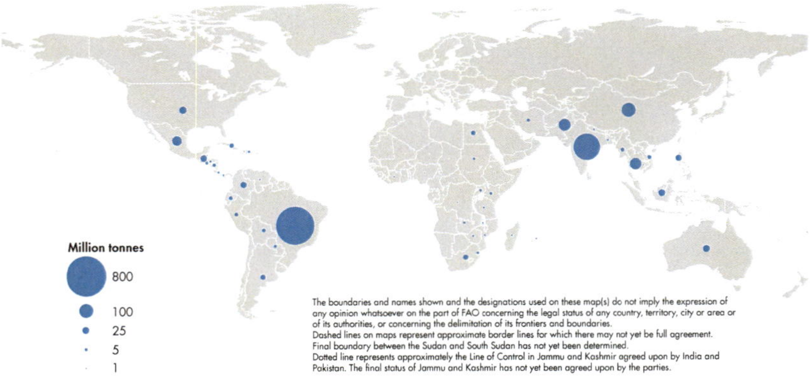

Source: FAO. 2022. FAOSTAT: Production: Crops and livestock products. In: *FAO*. Rome. Cited October 2022.
https://www.fao.org/faostat/en/#data/QCL based on UN Geospatial. 2020. Map geodata [shapefiles]. New York, USA, UN.
https://doi.org/10.4060/cc2211en-map10

FIGURE 15. **WORLD PRODUCTION OF MEAT, MAIN ITEMS**

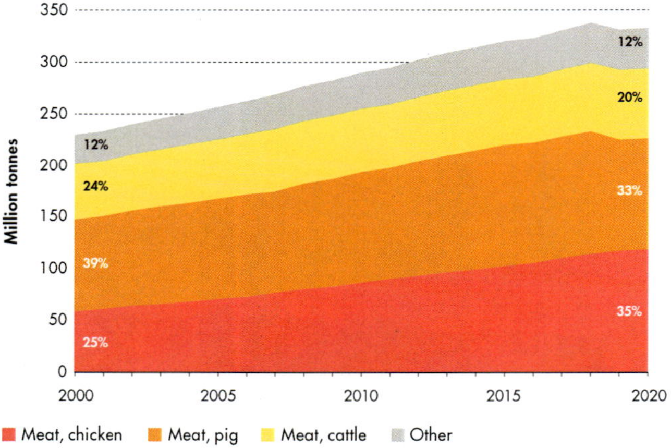

Note: Percentages on the figure indicate the shares in the total; they may not tally due to rounding.
Source: FAO. 2022. FAOSTAT: Production: Crops and livestock products. In: *FAO*. Rome. Cited October 2022.
https://www.fao.org/faostat/en/#data/QCL
https://doi.org/10.4060/cc2211en-fig26

MAP 15. **PRODUCTION OF CHICKEN MEAT (2020)**

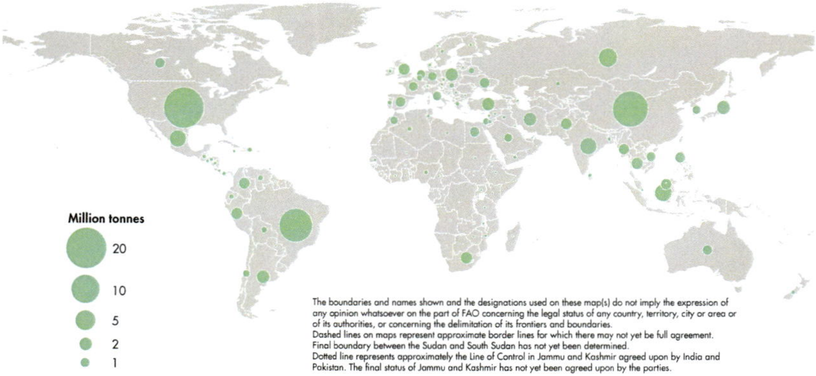

Source: FAO. 2022. FAOSTAT: Production: Crops and livestock products. In: *FAO*. Rome. Cited October 2022.
https://www.fao.org/faostat/en/#data/QCL based on UN Geospatial. 2020. Map geodata [shapefiles]. New York, USA, UN.
https://doi.org/10.4060/cc2211en-map16

FIGURE 16. **WORLD PRODUCTION OF MILK BY REGION**

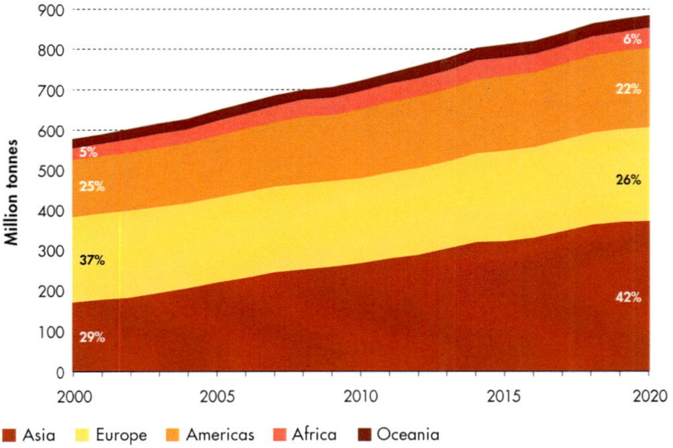

Note: Percentages on the figure indicate the shares in the total; they may not tally due to rounding.
Source: FAO. 2022. FAOSTAT: Production: Crops and livestock products. In: *FAO*. Rome. Cited October 2022.
https://www.fao.org/faostat/en/#data/QCL
https://doi.org/10.4060/cc2211en-fig28

MAP 16. **PRODUCTION OF MILK (2020)**

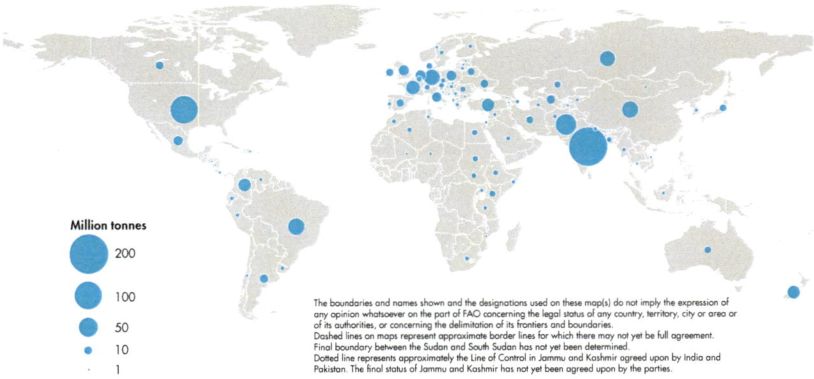

The boundaries and names shown and the designations used on these map(s) do not imply the expression of any opinion whatsoever on the part of FAO concerning the legal status of any country, territory, city or area or of its authorities, or concerning the delimitation of its frontiers and boundaries.
Dashed lines on maps represent approximate border lines for which there may not yet be full agreement.
Final boundary between the Sudan and South Sudan has not yet been determined.
Dotted line represents approximately the Line of Control in Jammu and Kashmir agreed upon by India and Pakistan. The final status of Jammu and Kashmir has not yet been agreed upon by the parties.

Source: FAO. 2022. FAOSTAT: Land use indicators. In: *FAO*. Rome. Cited October 2022.
https://www.fao.org/faostat/en/#data/QCL based on UN Geospatial. 2020. Map geodata [shapefiles]. New York, USA, UN.
https://doi.org/10.4060/cc2211en-map19

9 FISHERIES AND AQUACULTURE

FIGURE 17. **WORLD CAPTURE FISHERIES AND AQUACULTURE PRODUCTION BY SPECIES GROUP**

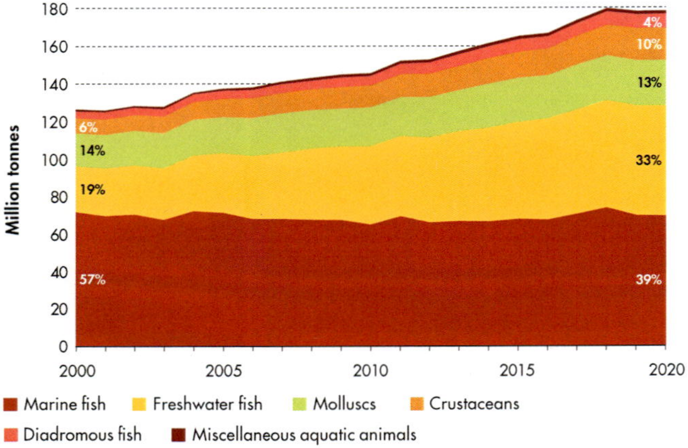

Note: Excludes aquatic mammals, crocodiles, alligators and caimans, pearls and shells, corals, sponges and algae. Percentages on the figure indicate the shares in the total; they may not tally due to rounding.
Source: FAO. 2022. Fisheries and Aquaculture: Global production by production source Quantity (1950 - 2020). In: *FAO*. Rome. Cited October 2022. https://www.fao.org/fishery/statistics-query/en/global_production/global_production_quantity
https://doi.org/10.4060/cc2211en-fig30

MAP 17. **CAPTURE FISHERIES AND AQUACULTURE PRODUCTION (2020)**

Note: Excludes aquatic mammals, crocodiles, alligators and caimans, pearls and shells, corals, sponges and algae.
Source: FAO. 2022. Fisheries and Aquaculture: Global production by production source Quantity (1950 - 2020). In: *FAO*. Rome. Cited October 2022. https://www.fao.org/fishery/statistics-query/en/global_production/global_production_quantity based on UN Geospatial. 2020. Map geodata [shapefiles]. New York, USA, UN.
https://doi.org/10.4060/cc2211en-map20

FIGURE 18. **WORLD CAPTURE FISHERIES AND AQUACULTURE PRODUCTION BY PRODUCTION MODE**

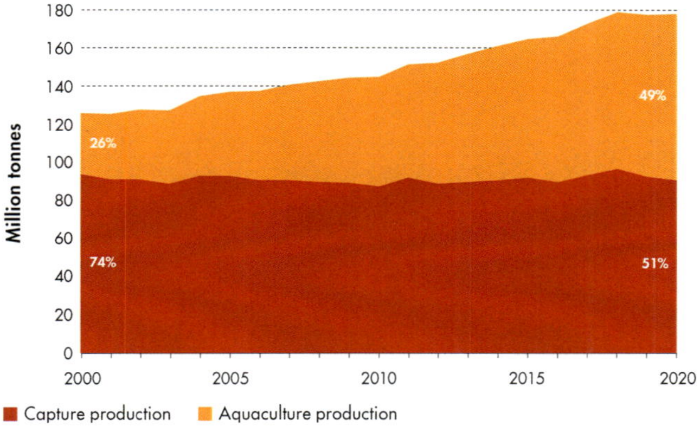

Note: Excludes aquatic mammals, crocodiles, alligators and caimans, pearls and shells, corals, sponges and algae. Percentages on the figure indicate the shares in the total; they may not tally due to rounding.
Source: FAO. 2022. Fisheries and Aquaculture: Global production by production source Quantity (1950 - 2020). In: *FAO*. Rome. Cited October 2022. https://www.fao.org/fishery/statistics-query/en/global_production/global_production_quantity
https://doi.org/10.4060/cc2211en-fig31

MAP 18. **AQUACULTURE PRODUCTION (2020)**

Note: Excludes aquatic mammals, crocodiles, alligators and caimans, pearls and shells, corals, sponges and algae.
Source: FAO. 2022. Fisheries and Aquaculture: Global production by production source Quantity (1950 - 2020).
In: *FAO*. Rome. Cited October 2022. https://www.fao.org/fishery/statistics-query/en/global_production/global_production_quantity based on UN Geospatial. 2020. Map geodata [shapefiles]. New York, USA, UN.
https://doi.org/10.4060/cc2211en-map21

10 FORESTRY

FIGURE 19. **WORLD PRODUCTION OF ROUNDWOOD BY TYPE**

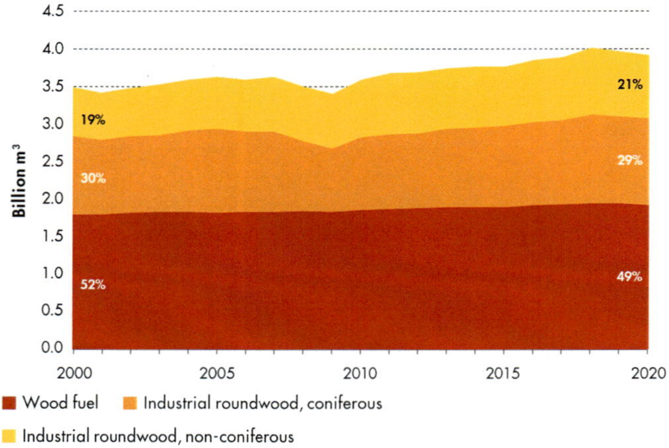

Note: Percentages on the figure indicate the shares in the total; they may not tally due to rounding.
Source: FAO. 2022. FAOSTAT: Forestry Production and Trade. In: *FAO*. Rome. Cited October 2022.
https://www.fao.org/faostat/en/#data/FO
https://doi.org/10.4060/cc2211en-fig33

MAP 19. **SHARE OF FOREST AREA IN LAND AREA (2020)**

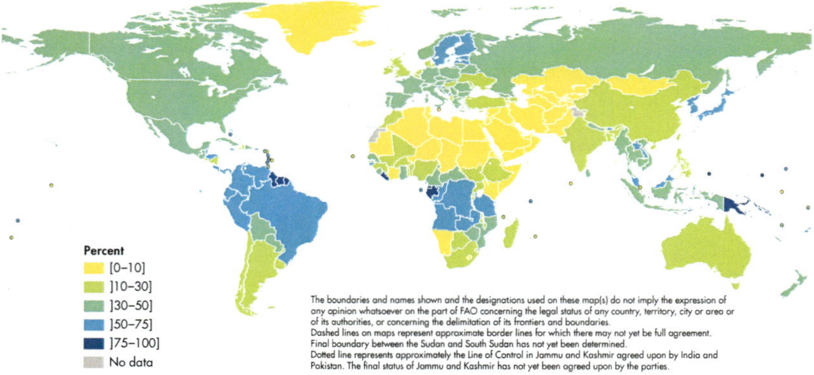

Source: FAO. 2022. FAOSTAT: Land use indicators. In: *FAO*. Rome. Cited October 2022.
http://www.fao.org/faostat/en/#data/EL based on UN Geospatial. 2020. Map geodata [shapefiles]. New York, USA, UN.
https://doi.org/10.4060/cc2211en-map29

FIGURE 20. **WORLD PRODUCTION OF SELECTED FOREST PRODUCTS**

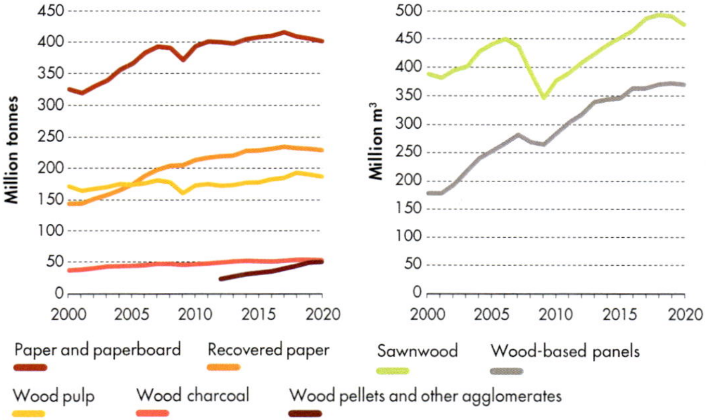

Source: FAO. 2022. FAOSTAT: Forestry Production and Trade. In: *FAO*. Rome. Cited October 2022.
https://www.fao.org/faostat/en/#data/FO
https://doi.org/10.4060/cc2211en-fig35

MAP 20. **IMPORTERS AND EXPORTERS OF FOREST PRODUCTS (2020)**

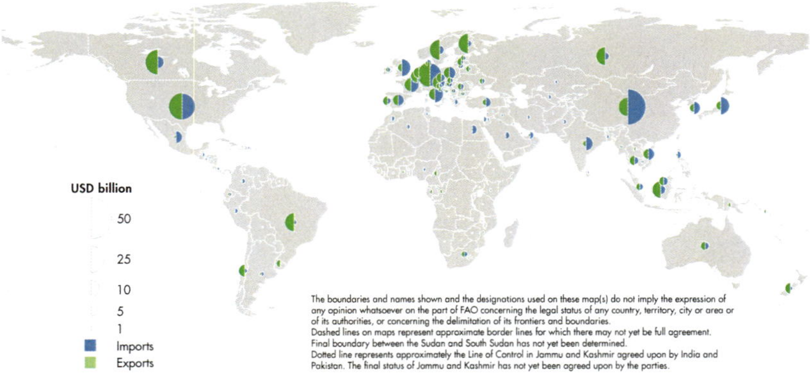

Source: FAO. 2022. FAOSTAT: Forestry Production and Trade. In: *FAO*. Rome. Cited October 2022.
https://www.fao.org/faostat/en/#data/FO based on UN Geospatial. 2020. Map geodata [shapefiles]. New York, USA, UN.
https://doi.org/10.4060/cc2211en-map22

FIGURE 21. **VALUE OF WORLD FOOD EXPORTS BY GROUP**

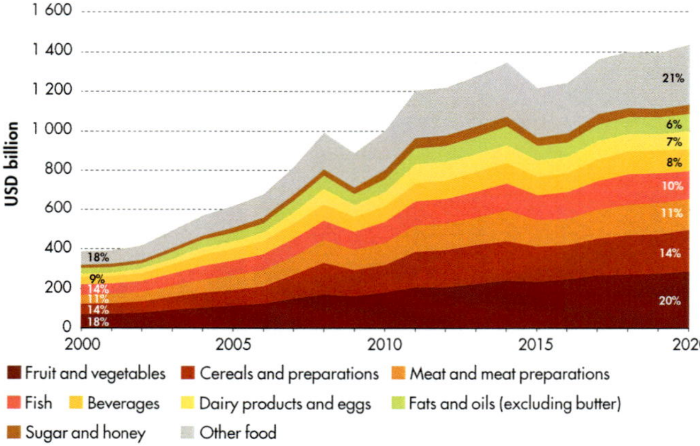

Note: Values for fish exclude trade of aquatic mammals, crocodiles, alligators and caimans, fishmeal, fish oil, ornamental fish, fish for culture and algae. Percentages on the figure indicate the shares in the total; they may not tally due to rounding.
Source: FAO. 2022. FAOSTAT: Trade: Crops and livestock products. In: *FAO*. Rome. Cited October 2022. https://www.fao.org/faostat/en/#data/TCL and FAO. 2022. Fisheries and Aquaculture: Global fish trade - All partners aggregated Value (1976 - 2020). In: FAO. Rome. Cited October 2022. https://www.fao.org/fishery/statistics-query/en/trade/trade_value https://doi.org/10.4060/cc2211en-fig36

MAP 21. **CEREALS EXPORTERS (2020)**

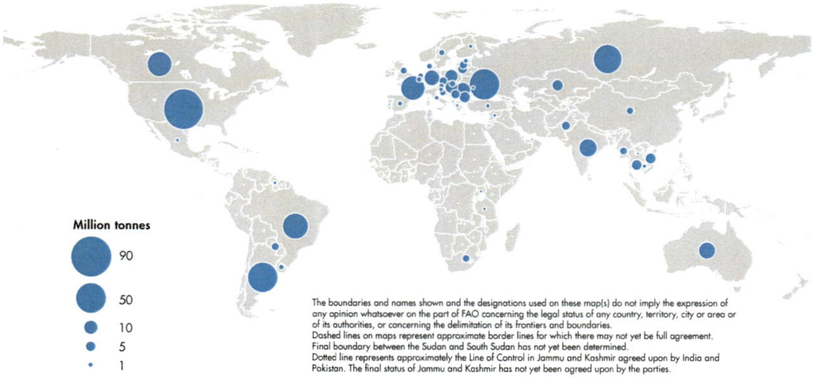

Source: FAO. 2022. Trade: Crops and livestock products. In: *FAO*. Rome. Cited October 2022. https://www.fao.org/faostat/en/#data/TCL based on UN Geospatial. 2020. Map geodata [shapefiles]. New York, USA, UN. https://doi.org/10.4060/cc2212en-map21

FIGURE 22. **FOOD NET TRADE BY REGION**

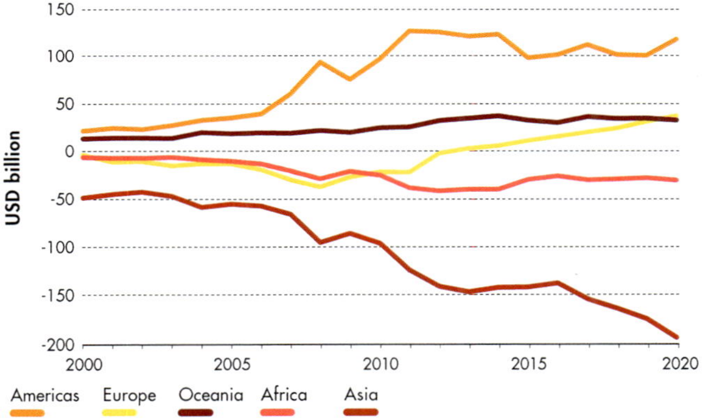

Note: Values for fish exclude trade of aquatic mammals, crocodiles, alligators and caimans, fishmeal, fish oil, ornamental fish, fish for culture and algae.
Source: FAO. 2022. FAOSTAT: Trade: Crops and livestock products. In: *FAO*. Rome. Cited October 2022. https://www.fao.org/faostat/en/#data/TCL and FAO. 2022. Fisheries and Aquaculture: Global fish trade - All partners aggregated Value (1976 - 2020). In: *FAO*. Rome. Cited October 2022. https://www.fao.org/fishery/statistics-query/en/trade/trade_value
https://doi.org/10.4060/cc2211en-fig37

MAP 22. **IMPORTERS AND EXPORTERS OF FOOD (2020)**

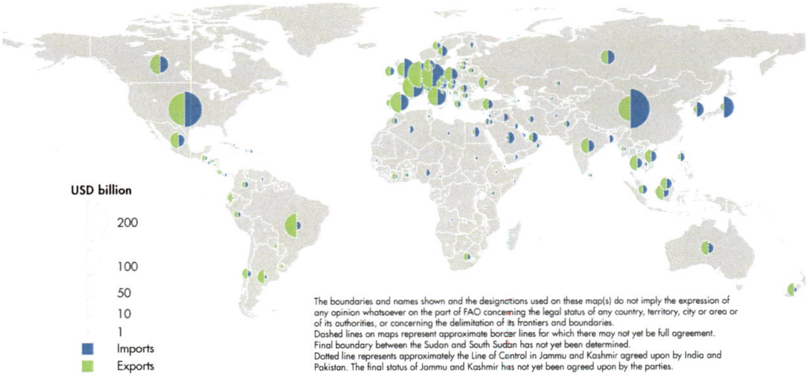

Note: See Figure 22.
Source: FAO. 2022. Trade: Crops and livestock products. In: *FAO*. Rome. Cited October 2022. https://www.fao.org/faostat/en/#data/TCL and FAO. 2022. Fisheries and Aquaculture: Global fish trade - All partners aggregated Value (1976 - 2020). In: *FAO*. Rome. Cited October 2022. https://www.fao.org/fishery/statistics-query/en/trade/trade_value based on UN Geospatial. 2020. Map geodata [shapefiles]. New York, USA. UN.
https://doi.org/10.4060/cc2211en-map23

12 PRICES

FIGURE 23. **INFLATION IN FOOD CONSUMER PRICES BY REGION**

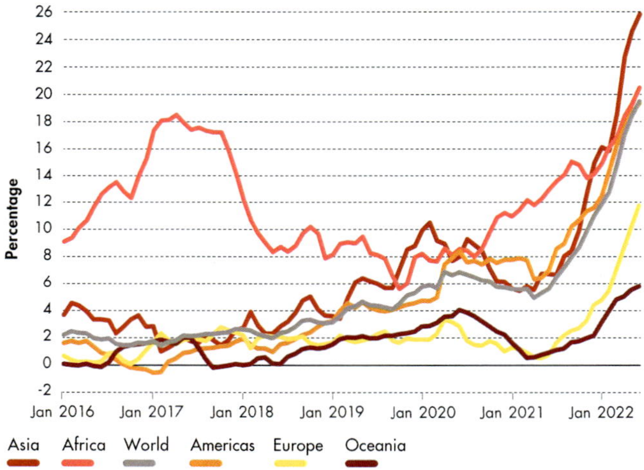

Source: FAO. 2022. FAOSTAT: Consumer Price Indices. In: *FAO*. Rome. Cited October 2022.
https://www.fao.org/faostat/en/#data/CP
https://doi.org/10.4060/cc2211en-fig46

MAP 23. **INFLATION IN FOOD CONSUMER PRICES (2021 AVERAGE)**

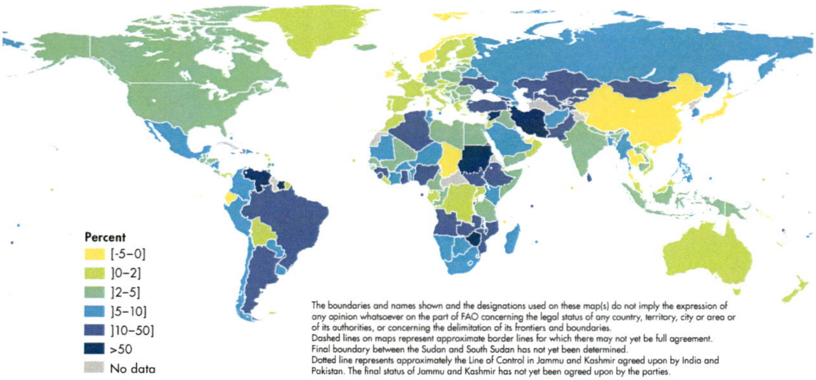

The boundaries and names shown and the designations used on these map(s) do not imply the expression of any opinion whatsoever on the part of FAO concerning the legal status of any country, territory, city or area or of its authorities, or concerning the delimitation of its frontiers and boundaries.
Dashed lines on maps represent approximate border lines for which there may not yet be full agreement.
Final boundary between the Sudan and South Sudan has not yet been determined.
Dotted line represents approximately the Line of Control in Jammu and Kashmir agreed upon by India and Pakistan. The final status of Jammu and Kashmir has not yet been agreed upon by the parties.

Source: FAO. 2022. FAOSTAT: Consumer Price Indices. In: *FAO*. Rome. Cited October 2022.
https://www.fao.org/faostat/en/#data/CP based on UN Geospatial. 2020. Map geodata [shapefiles]. New York, USA, UN.
https://doi.org/10.4060/cc2211en-map24

FIGURE 24. **ANNUAL CHANGES IN PRICES RECEIVED BY FARMERS, TOP AND BOTTOM COUNTRIES (2021)**

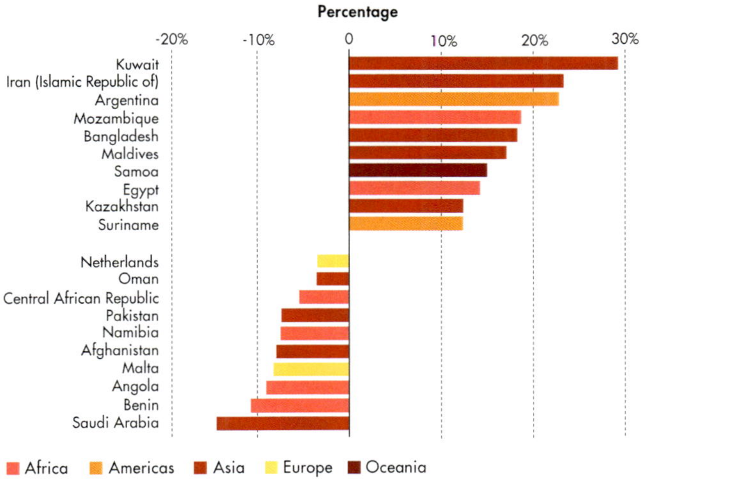

Source: FAO. 2022. FAOSTAT: Producer prices. In: *FAO*. Rome. Cited October 2022. http://www.fao.org/faostat/en/#data/PP
https://doi.org/10.4060/cc2211en-fig45

MAP 24. **ANNUAL CHANGES IN PRICES RECEIVED BY FARMERS (2020)**

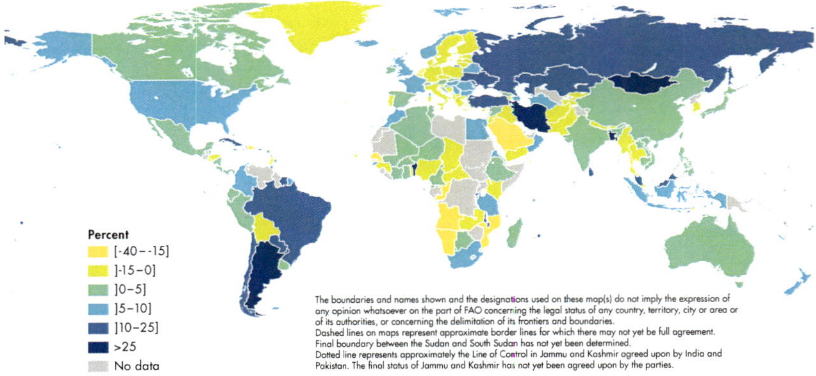

Source: FAO. 2022. FAOSTAT: Producer prices. In: *FAO*. Rome. Cited October 2022.
http://www.fao.org/faostat/en/#data/PP based on UN Geospatial. 2020. Map geodata [shapefiles]. New York, USA, UN.
https://doi.org/10.4060/cc2212en-map24

13 FOOD SUPPLY

FIGURE 25. **AVERAGE DIETARY ENERGY SUPPLY BY REGION**

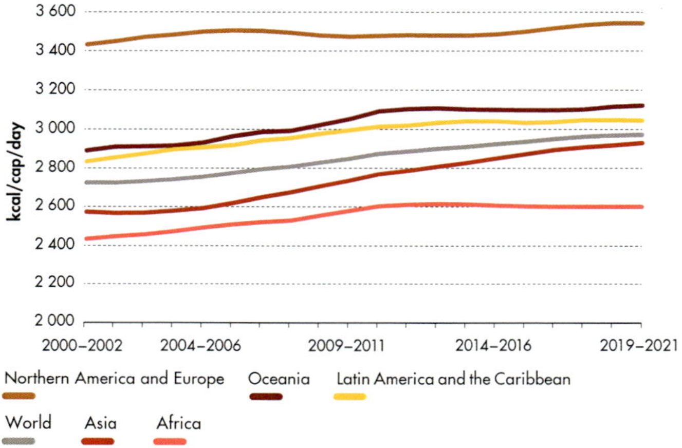

Note: This series is used for estimating the prevalence of undernourishment. Values for 2020 and 2021 are projections.
Source: FAO. 2022. FAOSTAT: Suite of Food Security Indicators. In: *FAO*. Rome. Cited October 2022.
https://www.fao.org/faostat/en/#data/FS
https://doi.org/10.4060/cc2211en-fig51

MAP 25. **AVERAGE DIETARY ENERGY SUPPLY (2019–2021 AVERAGE)**

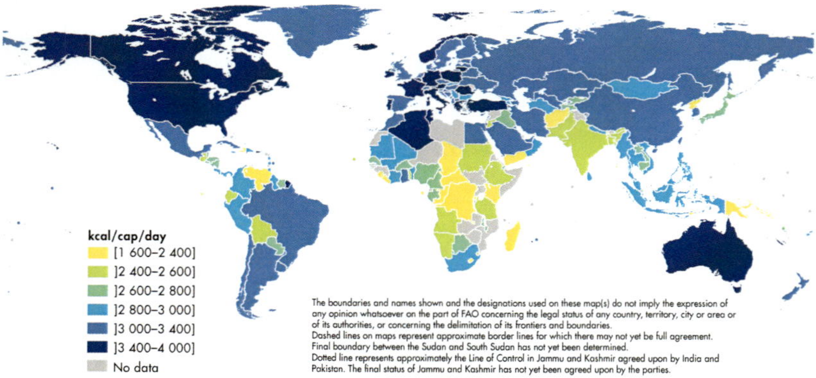

The boundaries and names shown and the designations used on these map(s) do not imply the expression of any opinion whatsoever on the part of FAO concerning the legal status of any country, territory, city or area or of its authorities, or concerning the delimitation of its frontiers and boundaries.
Dashed lines on maps represent approximate border lines for which there may not yet be full agreement.
Final boundary between the Sudan and South Sudan has not yet been determined.
Dotted line represents approximately the Line of Control in Jammu and Kashmir agreed upon by India and Pakistan. The final status of Jammu and Kashmir has not yet been agreed upon by the parties.

Note: This series is used for estimating the prevalence of undernourishment. Values for 2020 and 2021 are projections.
Source: FAO. 2022. FAOSTAT: Suite of Food Security Indicators. In: *FAO*. Rome. Cited October 2022.
https://www.fao.org/faostat/en/#data/FS based on UN Geospatial. 2020. Map geodata [shapefiles]. New York, USA, UN.
https://doi.org/10.4060/cc2212en-map25

FIGURE 26. **CEREAL IMPORT DEPENDENCY BY REGION**

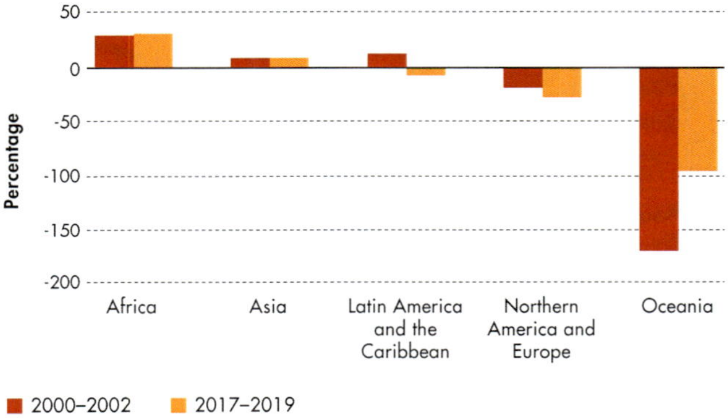

Source: FAO. 2022. FAOSTAT: Suite of Food Security Indicators. In: *FAO*. Rome. Cited October 2022.
https://www.fao.org/faostat/en/#data/FS
https://doi.org/10.4060/cc2212en-fig26

MAP 26. **CEREAL IMPORT DEPENDENCY (2017–2019 AVERAGE)**

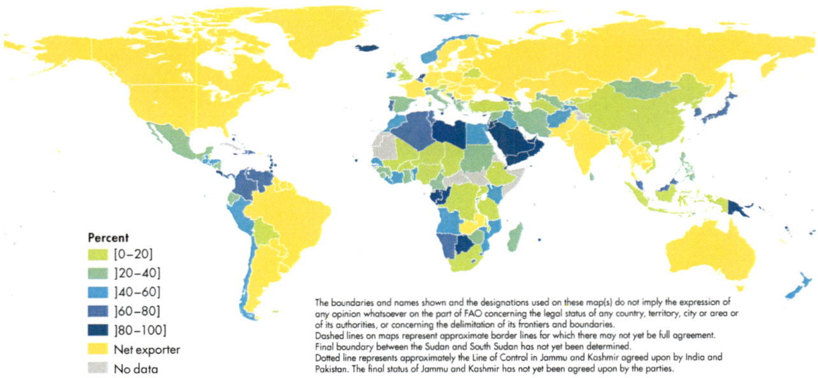

The boundaries and names shown and the designations used on these map(s) do not imply the expression of any opinion whatsoever on the part of FAO concerning the legal status of any country, territory, city or area or of its authorities, or concerning the delimitation of its frontiers and boundaries.
Dashed lines on maps represent approximate border lines for which there may not yet be full agreement.
Final boundary between the Sudan and South Sudan has not yet been determined.
Dotted line represents approximately the Line of Control in Jammu and Kashmir agreed upon by India and Pakistan. The final status of Jammu and Kashmir has not yet been agreed upon by the parties.

Source: FAO. 2022. FAOSTAT: Suite of Food Security Indicators. In: *FAO*. Rome. Cited October 2022.
https://www.fao.org/faostat/en/#data/FS based on UN Geospatial. 2020. Map geodata [shapefiles]. New York, USA, UN.
https://doi.org/10.4060/cc2212en-map26

FIGURE 27. **PREVALENCE OF UNDERNOURISHMENT BY REGION**

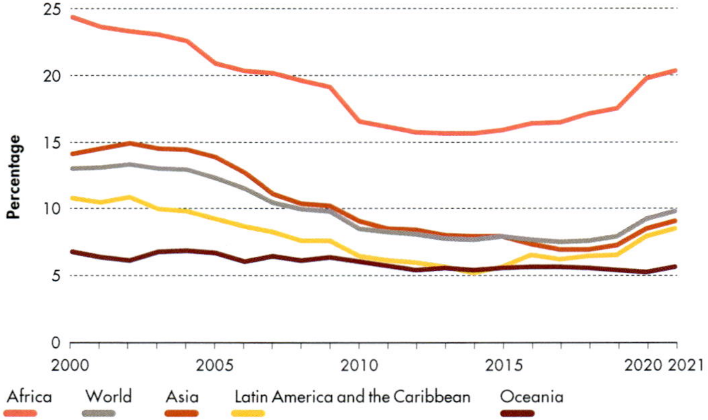

Note: The prevalence of undernourishment for Northern America and Europe is estimated to be less than 2.5 percent. Projected values for 2020 and 2021 are based on the middle of the projected range.
Source: FAO. 2022. FAOSTAT: Suite of Food Security Indicators. In: *FAO*. Rome. Cited October 2022.
https://www.fao.org/faostat/en/#data/FS
https://doi.org/10.4060/cc2211en-fig47

MAP 27. **PREVALENCE OF UNDERNOURISHMENT (2019–2021 AVERAGE)**

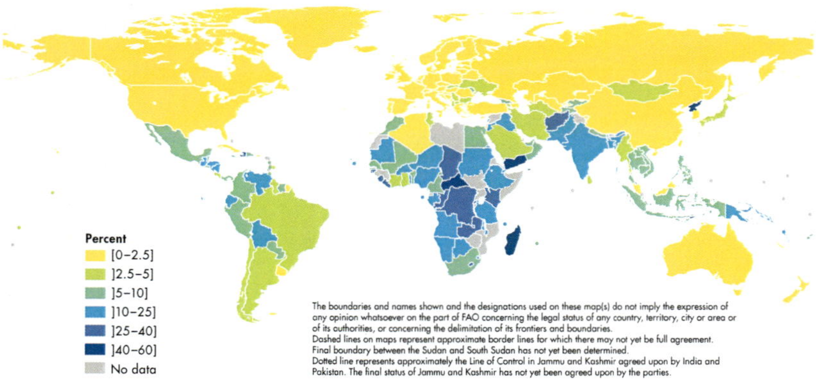

Note: Projected values for 2020 and 2021 are based on the middle of the projected range.
Source: FAO. 2022. FAOSTAT: Suite of Food Security Indicators. In: *FAO*. Rome. Cited October 2022.
https://www.fao.org/faostat/en/#data/FS based on UN Geospatial. 2020. Map geodata [shapefiles]. New York, USA, UN.
https://doi.org/10.4060/cc2211en-map25

FIGURE 28. **FOOD INSECURITY LEVELS BASED ON THE FOOD INSECURITY EXPERIENCE SCALE BY REGION**

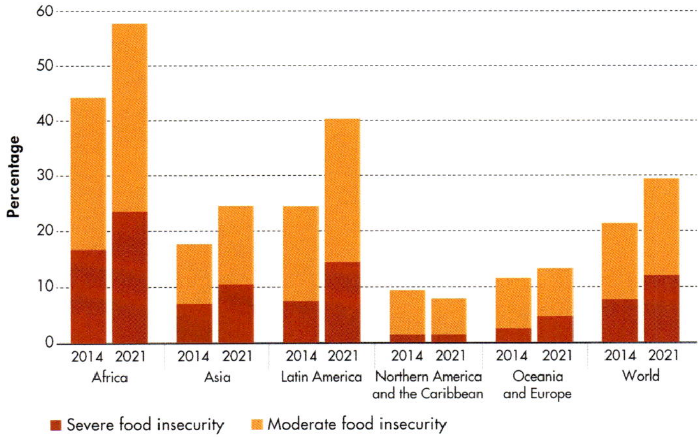

Source: FAO. 2022. FAOSTAT: Suite of Food Security Indicators. In: *FAO*. Rome. Cited October 2022. https://www.fao.org/faostat/en/#data/FS
https://doi.org/10.4060/cc2211en-fig49

MAP 28. **NUMBER OF UNDERNOURISHED PEOPLE (2019–2021)**

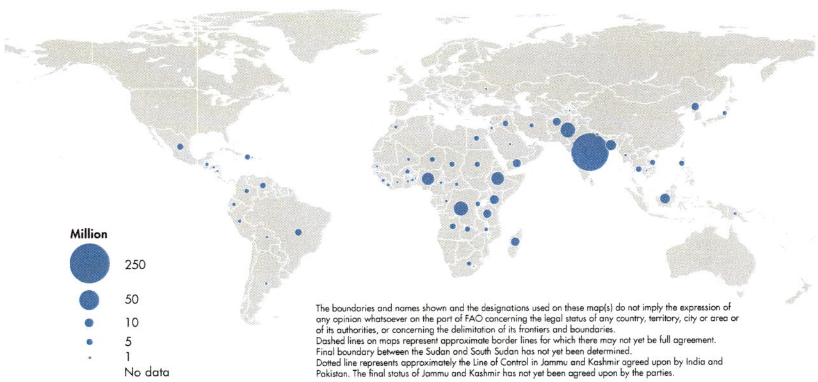

Note: Projected values for 2020 and 2021 are based on the middle of the projected range.
Source: FAO. 2022. FAOSTAT: Suite of Food Security Indicators. In: *FAO*. Rome. Cited October 2022. https://www.fao.org/faostat/en/#data/FS based on UN Geospatial. 2020. Map geodata [shapefiles]. New York, USA, UN.
https://doi.org/10.4060/cc2211en-map26

15 NUTRITION

FIGURE 29. **PREVALENCE OF STUNTING IN CHILDREN UNDER 5 YEARS BY REGION**

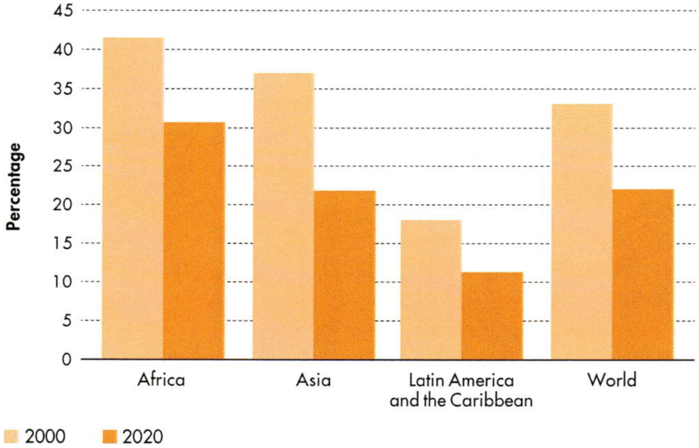

Source: FAO. 2022. FAOSTAT: Suite of Food Security Indicators. In: *FAO*. Rome. Cited October 2022.
https://www.fao.org/faostat/en/#data/FS
https://doi.org/10.4060/cc2211en-fig56

MAP 29. **PREVALENCE OF STUNTING IN CHILDREN UNDER 5 YEARS (2020)**

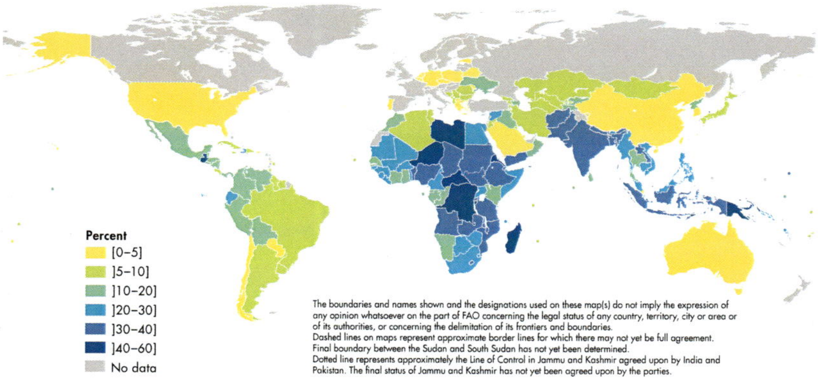

Source: FAO. 2022. FAOSTAT: Suite of Food Security Indicators. In: *FAO*. Rome. Cited October 2022.
https://www.fao.org/faostat/en/#data/FS based on UN Geospatial. 2020. Map geodata [shapefiles]. New York, USA, UN.
https://doi.org/10.4060/cc2211en-map27

FIGURE 30. **PREVALENCE OF OBESITY IN THE ADULT POPULATION BY REGION**

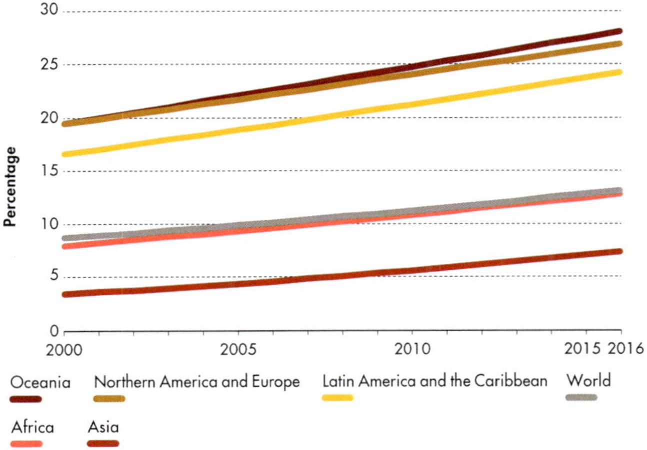

Source: FAO. 2022. FAOSTAT: Suite of Food Security Indicators. In: *FAO*. Rome. Cited October 2022.
https://www.fao.org/faostat/en/#data/FS
https://doi.org/10.4060/cc2211en-fig57

MAP 30. **PREVALENCE OF OBESITY IN THE ADULT POPULATION (2016)**

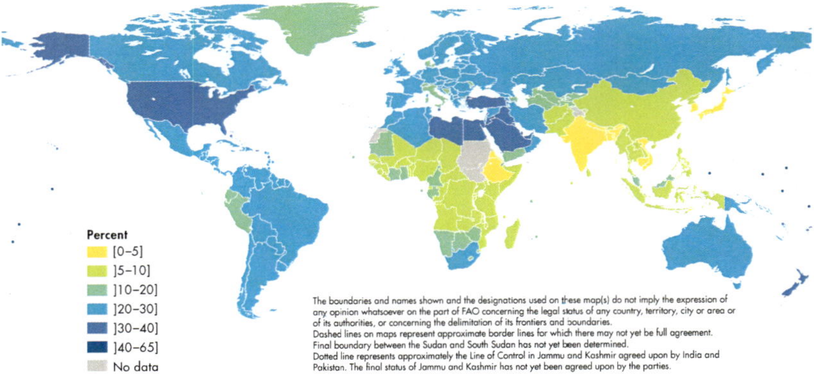

Source: FAO. 2022. FAOSTAT: Suite of Food Security Indicators. In: *FAO*. Rome. Cited October 2022.
https://www.fao.org/faostat/en/#data/FS based on UN Geospatial. 2020. Map geodata [shapefiles]. New York, USA, UN.
https://doi.org/10.4060/cc2211en-map28

FIGURE 31. **AREA EQUIPPED FOR IRRIGATION BY REGION**

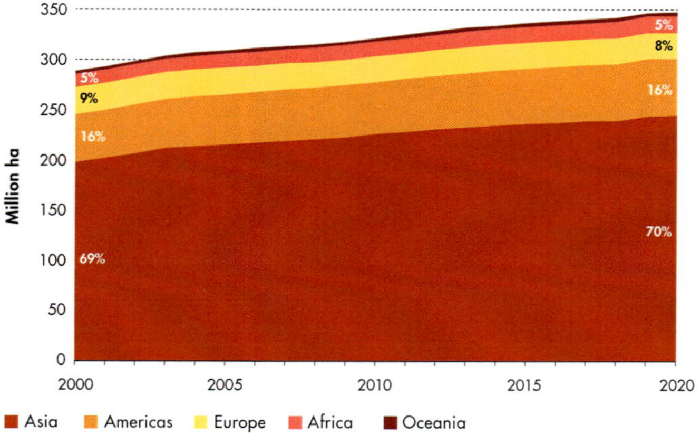

Note: Percentages on the figure indicate the shares in the total; they may not tally due to rounding.
Source: FAO. 2022. FAOSTAT: Land Use. In: *FAO*. Rome. Cited October 2022. http://www.fao.org/faostat/en/#data/RL
https://doi.org/10.4060/cc2211en-fig07

MAP 31. **TOTAL RENEWABLE WATER RESOURCES PER CAPITA (2019)**

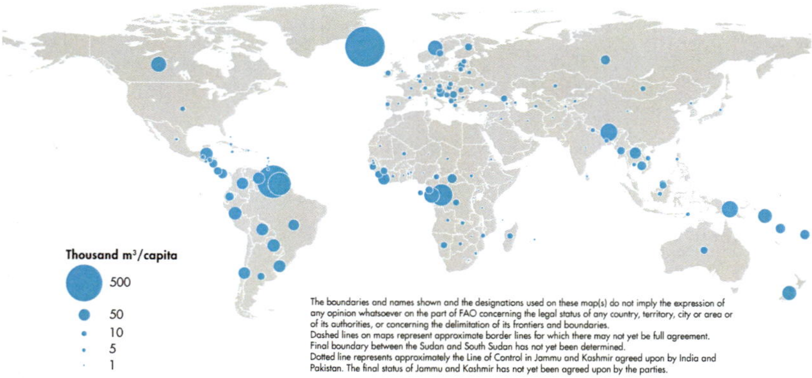

Source: FAO. 2022. AQUASTAT. In: *FAO*. Rome. Cited October 2022.
https://www.fao.org/aquastat/statistics/query/index.html?lang=en based on UN Geospatial. 2020. Map geodata [shapefiles]. New York, USA, UN.
https://doi.org/10.4060/cc2211en-map30

FIGURE 32. **WATER STRESS, TOP COUNTRIES (2019)**

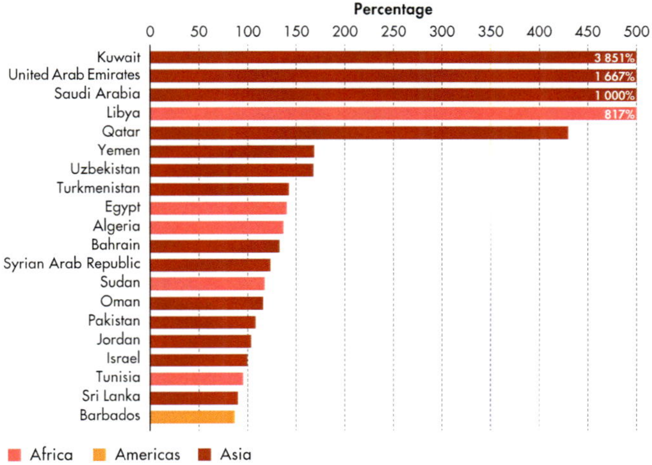

Source: FAO. 2022. AQUASTAT. In: *FAO*. Rome. Cited October 2022.
https://www.fao.org/aquastat/statistics/query/index.html?lang=en
https://doi.org/10.4060/cc2211en-fig64

MAP 32. **WATER STRESS (2019)**

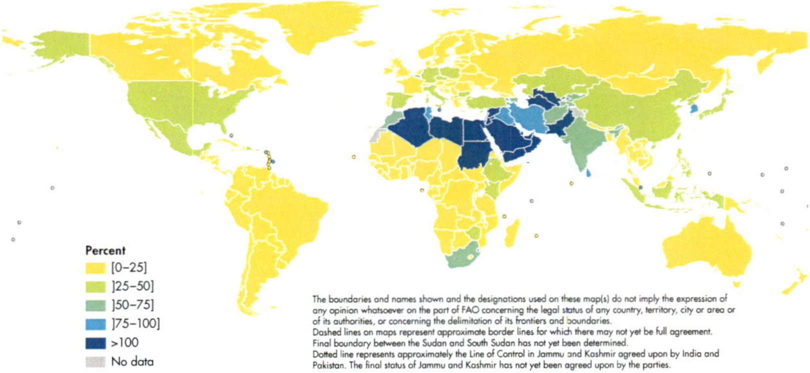

Source: FAO. 2022. AQUASTAT. In: *FAO*. Rome. Cited October 2022.
https://www.fao.org/aquastat/statistics/query/index.html?lang=en based on UN Geospatial. 2020. Map geodata [shapefiles]. New York, USA, UN.
https://doi.org/10.4060/cc2211en-map31

FIGURE 33. **WORLD GREENHOUSE GAS EMISSIONS ON AGRICULTURAL LAND**

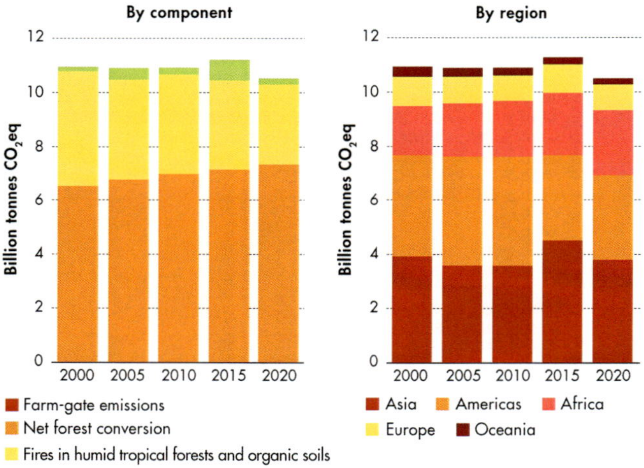

Source: FAO. 2022. FAOSTAT: Emissions Totals. In: *FAO*. Rome. Cited October 2022.
https://www.fao.org/faostat/en/#data/GT
https://doi.org/10.4060/cc2211en-fig66

MAP 33. **GREENHOUSE GAS EMISSIONS ON AGRICULTURAL LAND (2020)**

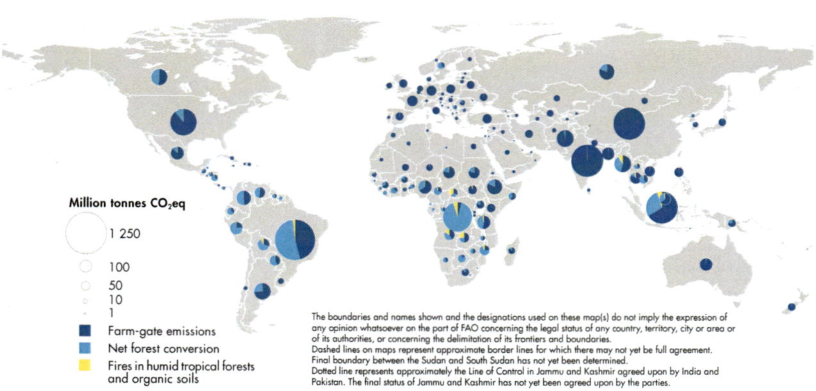

Source: FAO. 2022. FAOSTAT: Emissions Totals. In: *FAO*. Rome. Cited October 2022.
https://www.fao.org/faostat/en/#data/GT based on UN Geospatial. 2020. Map geodata [shapefiles]. New York, USA, UN.
https://doi.org/10.4060/cc2211en-map32

FIGURE 34. SHARE OF GREENHOUSE GAS EMISSIONS ON AGRICULTURAL AND IN TOTAL EMISSIONS BY REGION

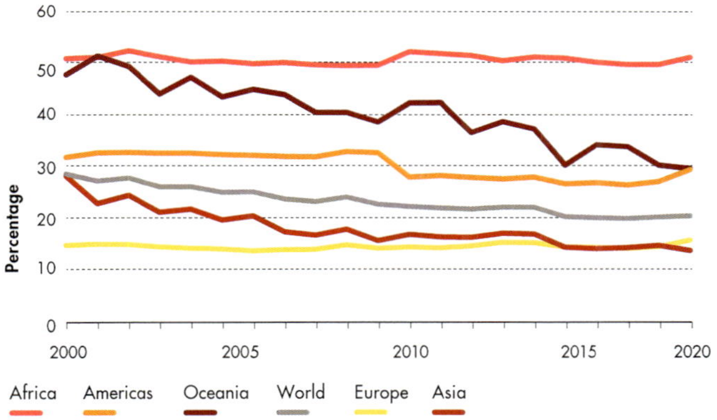

Note: Emissions are calculated using the methodology from the IPCC's Fifth Assessment Report (AR5).
Source: *FAO*. 2022. FAOSTAT: Emissions shares. In: *FAO*. Rome. Cited October 2022.
https://www.fao.org/faostat/en/#data/EM
https://doi.org/10.4060/cc2212en-fig34

MAP 34. SHARE OF GREENHOUSE GAS EMISSIONS ON AGRICULTURAL LAND IN TOTAL EMISSIONS (2020)

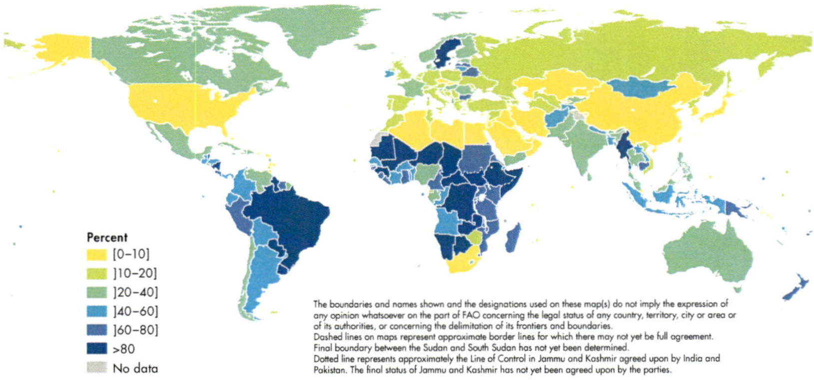

Note: Emissions are calculated using the methodology from the IPCC's Fifth Assessment Report (AR5).
Source: FAO. 2022. FAOSTAT: Emissions shares. In: *FAO*. Rome. Cited October 2022.
https://www.fao.org/faostat/en/#data/EM based on UN Geospatial. 2020. Map geodata [shapefiles]. New York, USA, UN.
https://doi.org/10.4060/cc2212en-map34

DATA TABLES

SELECTED INDICATORS – SOCIOECONOMIC

	POPULATION (PROJECTION)	GDP (CURRENT PRICES)	AGRICULTURE, FORESTRY AND FISHING VALUE ADDED SHARE IN GDP (2015 PRICES)	EMPLOYMENT IN AGRICULTURE, FORESTRY AND FISHING	SHARE OF EMPLOYMENT IN AGRICULTURE, FORESTRY AND FISHING IN TOTAL EMPLOYMENT
	MILLION, 2020	USD BILLION, 2020	PERCENT, 2020	THOUSAND, 2020	PERCENT, 2020
WORLD	7 841.0	84 659.9	4.4	866 289.0	26.6
AFRICA	1 360.7	2 414.6	16.2	225 595.4	48.4
AMERICAS	1 025.8	27 148.0	1.9	42 934.6	9.6
ASIA	4 664.3	32 402.5	7.4	578 944.0	29.2
EUROPE	746.2	21 010.5	1.7	17 545.9	5.1
OCEANIA	43.9	1 684.4	2.7	1 269.0	6.5
Afghanistan	39.0	19.8	30.6	3 945.9	46.9
Albania	2.9	14.9	19.0	425.6	34.6
Algeria	43.5	147.7	12.6	1 117.4	10.3
Andorra	0.1	2.9	0.5		
Angola	33.4	62.3	10.5	7 496.1	58.7
Antigua and Barbuda	0.1	1.4	2.0		
Argentina	45.0	383.1	5.6	1 422.3	7.7
Armenia	2.8	12.6	11.6	292.4	30.3
Australia	25.7	1 423.5	2.1	314.8	2.4
Austria	8.9	433.3	1.2	164.5	3.7
Azerbaijan	10.3	42.6	7.9	1 595.0	34.2
Bahamas	0.4	9.9	0.4	6.0	3.2
Bahrain	1.5	33.9	0.3	9.7	1.0
Bangladesh	167.4	329.5	12.4	24 541.0	37.1
Barbados	0.3	4.4	1.7	3.6	2.8
Belarus	9.6	60.3	6.9	383.5	8.1
Belgium	11.6	521.9	0.6	45.9	0.9
Belize	0.4	1.6	11.7	34.5	21.2

GROSS FIXED CAPITAL FORMATION (AGRICULTURE, FORESTRY AND FISHING) AS A SHARE OF VALUE ADDED	PESTICIDE USE, TOTAL	FERTILIZERS USE, TOTAL	INFLATION IN FOOD CONSUMER PRICES	CHANGE IN PRICES RECEIVED BY FARMERS	
PERCENT, 2021	THOUSAND TONNES, 2020	THOUSAND TONNES, 2020	PERCENT, 2021	PERCENT, 2021	
14.0	2 661.1	200 570.5	7.0		**WORLD**
6.1	105.8	7 092.6	13.1		**AFRICA**
14.3	1 358.0	55 071.1	8.7		**AMERICAS**
12.7	658.5	110 465.0	8.0		**ASIA**
30.3	468.4	24 404.3	1.9		**EUROPE**
28.3	70.4	3 537.5	1.3		**OCEANIA**
7.6		68.2	5.5	-7.9	**Afghanistan**
14.1	0.8	60.8	3.9	1.6	**Albania**
14.4	6.1	155.7	10.0	4.8	**Algeria**
11.5					**Andorra**
6.0	0.0	42.1	32.4	-9.0	**Angola**
3.7	0.0	0.0	1.4		**Antigua and Barbuda**
12.3	241.3	2 317.2	49.6	22.9	**Argentina**
11.6	0.6	90.3	11.3	3.5	**Armenia**
34.2	63.4	2 583.1	1.2	3.3	**Australia**
54.8	5.6	169.7	0.8	1.5	**Austria**
10.4	0.5		8.1	3.6	**Azerbaijan**
37.2	0.2	1.9	2.1		**Bahamas**
9.0	0.0	2.0	-0.1		**Bahrain**
10.0	15.5	2 567.3	5.3	18.3	**Bangladesh**
1.9	0.2	0.4	3.3	3.3	**Barbados**
37.7	4.6	1 022.9	9.6	8.6	**Belarus**
76.4	5.6	255.4	-0.4	-3.4	**Belgium**
5.3	1.3	39.5	4.8	2.4	**Belize**

SELECTED INDICATORS – SOCIOECONOMIC (CONTINUED)

	POPULATION (PROJECTION)	GDP (CURRENT PRICES)	AGRICULTURE, FORESTRY AND FISHING VALUE ADDED SHARE IN GDP (2015 PRICES)	EMPLOYMENT IN AGRICULTURE, FORESTRY AND FISHING	SHARE OF EMPLOYMENT IN AGRICULTURE, FORESTRY AND FISHING IN TOTAL EMPLOYMENT
	MILLION, 2020	USD BILLION, 2020	PERCENT, 2020	THOUSAND, 2020	PERCENT, 2020
Benin	12.6	15.2	29.3	1 426.3	28.1
Bhutan	0.8	2.5	13.8	189.6	56.0
Bolivia (Plurinational State of)	11.9	36.6	12.8	1 688.2	29.2
Bosnia and Herzegovina	3.3	19.8	6.3	112.8	11.3
Botswana	2.5	15.8	2.2	168.7	23.1
Brazil	213.2	1 444.7	5.0	8 246.2	9.7
Brunei Darussalam	0.4	12.0	1.1	2.6	1.3
Bulgaria	7.0	69.9	4.3	194.9	6.3
Burkina Faso	21.5	17.4	22.0	5 440.9	73.3
Burundi	12.2	3.4	35.3	4 439.9	85.8
Cabo Verde	0.6	1.7	5.1	20.5	11.0
Cambodia	16.4	25.3	21.7	3 613.7	38.9
Cameroon	26.5	39.9	15.2	4 896.3	42.6
Canada	37.9	1 644.0	1.9	258.8	1.3
Central African Republic	5.3	2.3	32.2	1 278.2	68.5
Chad	16.6	11.2	39.7	3 589.6	68.9
Chile	19.3	252.9	3.7	508.5	6.6
China	1 456.9	15 096.6	7.6	188 740.2	24.0
Colombia	50.9	271.3	6.8	3 535.7	15.9
Comoros	0.8	1.2	38.7	74.7	35.0
Congo	5.7	10.1	9.1	618.6	36.3
Costa Rica	5.1	61.5	4.8	340.1	17.1

GROSS FIXED CAPITAL FORMATION (AGRICULTURE, FORESTRY AND FISHING) AS A SHARE OF VALUE ADDED	PESTICIDE USE, TOTAL	FERTILIZERS USE, TOTAL	INFLATION IN FOOD CONSUMER PRICES	CHANGE IN PRICES RECEIVED BY FARMERS	
PERCENT, 2021	THOUSAND TONNES, 2020	THOUSAND TONNES, 2020	PERCENT, 2021	PERCENT, 2021	
7.6		112.6	5.8	-10.7	**Benin**
13.0	0.0	1.3	9.2	0.1	**Bhutan**
12.6	19.3	48.9	0.6	0.4	**Bolivia (Plurinational State of)**
15.5	2.7	91.5	3.5	2.3	**Bosnia and Herzegovina**
8.3	0.0	13.3	5.9	2.1	**Botswana**
5.6	377.2	20 366.6	12.4	10.9	**Brazil**
17.5	0.0	0.7	2.4	2.2	**Brunei Darussalam**
27.8	4.2	486.2	2.7	4.8	**Bulgaria**
8.6	0.8	103.1	6.8	2.1	**Burkina Faso**
2.9	0.1	24.2	10.4		**Burundi**
14.5	0.0		0.5	0.8	**Cabo Verde**
9.8		181.0	2.7	2.2	**Cambodia**
8.5	7.3	83.0	4.2	2.9	**Cameroon**
16.0	78.9	5 061.0	2.2	2.2	**Canada**
6.1	0.0	0.3		-5.4	**Central African Republic**
2.8	0.0		-1.4	-0.7	**Chad**
14.7	9.8	490.4	5.4	1.3	**Chile**
14.6	273.4	45 796.1	-4.3	5.8	**China**
7.3	36.7	1 251.5	9.6	10.0	**Colombia**
5.8	0.0		-2.6		**Comoros**
14.0	0.0	5.4	3.0	3.8	**Congo**
15.2	14.1	160.5	2.2	1.1	**Costa Rica**

SELECTED INDICATORS – SOCIOECONOMIC (CONTINUED)

	POPULATION (PROJECTION)	GDP (CURRENT PRICES)	AGRICULTURE, FORESTRY AND FISHING VALUE ADDED SHARE IN GDP (2015 PRICES)	EMPLOYMENT IN AGRICULTURE, FORESTRY AND FISHING	SHARE OF EMPLOYMENT IN AGRICULTURE, FORESTRY AND FISHING IN TOTAL EMPLOYMENT
	MILLION, 2020	USD BILLION, 2020	PERCENT, 2020	THOUSAND, 2020	PERCENT, 2020
Côte d'Ivoire	26.8	61.1	15.4	3 828.3	45.0
Croatia	4.1	57.2	3.3	113.0	6.8
Cuba	11.3	107.4	3.8	890.4	17.7
Cyprus	1.2	24.6	1.8	17.0	2.8
Czechia	10.5	245.3	2.4	133.4	2.5
Democratic People's Republic of Korea	25.9	15.8	21.9	7 176.6	43.5
Democratic Republic of the Congo	92.9	45.3	17.3	17 100.2	55.3
Denmark	5.8	356.1	1.1	58.9	2.0
Djibouti	1.1	3.4	1.4	1.9	1.2
Dominica	0.1	0.5	12.9		
Dominican Republic	11.0	78.8	5.9	375.7	8.3
Ecuador	17.6	98.8	10.9	2 543.5	32.2
Egypt	107.5	369.3	10.7	5 117.4	19.8
El Salvador	6.3	24.6	5.6	390.6	15.2
Equatorial Guinea	1.6	10.0	2.5	253.6	55.5
Eritrea	3.6	2.1	17.6	942.1	62.4
Estonia	1.3	30.7	2.1	17.7	2.7
Eswatini	1.2	3.8	8.7	33.7	12.4
Ethiopia	117.2	96.6	30.5	34 270.9	63.7
Fiji	0.9	4.5	9.3	99.5	28.9
Finland	5.5	269.8	2.3	107.0	4.1

GROSS FIXED CAPITAL FORMATION (AGRICULTURE, FORESTRY AND FISHING) AS A SHARE OF VALUE ADDED	PESTICIDE USE, TOTAL	FERTILIZERS USE, TOTAL	INFLATION IN FOOD CONSUMER PRICES	CHANGE IN PRICES RECEIVED BY FARMERS	
PERCENT, 2021	THOUSAND TONNES, 2020	THOUSAND TONNES, 2020	PERCENT, 2021	PERCENT, 2021	
17.3	0.1	181.6	7.6	2.6	**Côte d'Ivoire**
26.0	1.6	177.6	1.6	2.4	**Croatia**
8.6		74.4		6.6	**Cuba**
5.7	1.2	16.7	-0.4	1.7	**Cyprus**
28.0	3.5	358.3	0.8	0.0	**Czechia**
		31.9			**Democratic People's Republic of Korea**
8.2		28.6	1.5		**Democratic Republic of the Congo**
35.7	3.2	342.9	0.6	5.5	**Denmark**
6.4			1.7		**Djibouti**
6.7		0.4	1.2		**Dominica**
3.4	7.1	172.4	9.9	9.7	**Dominican Republic**
4.1	34.1	376.7	-0.8	-2.0	**Ecuador**
4.2	11.4	1 593.1	4.5	14.2	**Egypt**
15.0	3.1	58.4	2.2	1.8	**El Salvador**
6.1			-1.3	-1.1	**Equatorial Guinea**
5.8	0.0	3.5		5.5	**Eritrea**
57.4	0.6	65.8	1.8	-1.0	**Estonia**
14.2			2.5		**Eswatini**
10.1	4.1	586.2	31.1	2.7	**Ethiopia**
7.0	1.2	7.5	6.5	0.4	**Fiji**
27.4	4.9	208.1	0.6	4.5	**Finland**

SELECTED INDICATORS – SOCIOECONOMIC (CONTINUED)

	POPULATION (PROJECTION)	GDP (CURRENT PRICES)	AGRICULTURE, FORESTRY AND FISHING VALUE ADDED SHARE IN GDP (2015 PRICES)	EMPLOYMENT IN AGRICULTURE, FORESTRY AND FISHING	SHARE OF EMPLOYMENT IN AGRICULTURE, FORESTRY AND FISHING IN TOTAL EMPLOYMENT
	MILLION, 2020	USD BILLION, 2020	PERCENT, 2020	THOUSAND, 2020	PERCENT, 2020
France	64.5	2 630.3	1.6	712.5	2.5
Gabon	2.3	15.1	6.5	155.3	29.0
Gambia	2.6	1.8	20.6	345.5	48.5
Georgia	3.8	15.9	7.5	674.5	40.4
Germany	83.3	3 846.4	0.7	534.0	1.3
Ghana	32.2	68.5	20.2	5 156.7	39.5
Greece	10.5	188.8	4.2	438.9	11.4
Grenada	0.1	1.0	4.9		
Guatemala	17.4	77.6	10.2	2 003.0	29.2
Guinea	13.2	15.5	19.1	2 662.7	59.2
Guinea-Bissau	2.0	1.3	46.3	390.7	50.3
Guyana	0.8	5.5	16.6	32.7	13.1
Haiti	11.3	15.5	17.0	1 947.0	45.6
Honduras	10.1	23.8	12.7	963.6	24.8
Hungary	9.8	155.8	3.3	205.6	4.4
Iceland	0.4	21.7	5.4	7.0	4.0
India	1 396.4	2 664.7	17.3	194 797.0	44.0
Indonesia	271.9	1 058.4	13.3	38 559.6	29.0
Iran (Islamic Republic of)	87.3	939.3	11.5	3 800.2	16.3
Iraq	42.6	166.8	7.2	1 822.7	19.8
Ireland	4.9	425.9	0.9	103.8	4.5
Israel	8.8	407.1	1.1	33.3	0.9
Italy	59.5	1 888.7	2.0	931.4	4.1

GROSS FIXED CAPITAL FORMATION (AGRICULTURE, FORESTRY AND FISHING) AS A SHARE OF VALUE ADDED	PESTICIDE USE, TOTAL	FERTILIZERS USE, TOTAL	INFLATION IN FOOD CONSUMER PRICES	CHANGE IN PRICES RECEIVED BY FARMERS	
PERCENT, 2021	THOUSAND TONNES, 2020	THOUSAND TONNES, 2020	PERCENT, 2021	PERCENT, 2021	
31.3	65.2	3 034.5	0.6	2.5	**France**
8.8		6.9	1.3		**Gabon**
9.2	0.6	0.3	9.6	8.2	**Gambia**
7.5	2.9	55.2	10.7	6.8	**Georgia**
48.3	48.0	1 904.1	3.2	2.1	**Germany**
11.5	9.7	270.0	10.3	1.6	**Ghana**
29.1	10.5	319.7	1.4	-2.0	**Greece**
9.4			1.7	0.0	**Grenada**
11.7	11.8	368.4	4.9		**Guatemala**
11.1	0.1	21.3	15.8	-0.8	**Guinea**
2.9	0.1		5.0	4.2	**Guinea-Bissau**
6.2	0.5	22.1			**Guyana**
6.1	0.0		19.6		**Haiti**
12.2	8.2	160.3	2.9	0.2	**Honduras**
39.9	8.7	654.1	3.4	3.5	**Hungary**
19.5	0.0	15.7	3.5	4.4	**Iceland**
13.8	61.7	32 535.6	3.7	2.2	**India**
15.3	1.6	6 527.7	2.7	1.8	**Indonesia**
5.3	5.9	1 072.1	59.1	23.4	**Iran (Islamic Republic of)**
5.1	0.3	274.2	4.8	0.1	**Iraq**
35.3	3.0	694.1	-0.3	0.4	**Ireland**
21.1	7.0		1.0	2.7	**Israel**
29.4	56.6	930.8	0.6	1.3	**Italy**

SELECTED INDICATORS – SOCIOECONOMIC (CONTINUED)

	POPULATION (PROJECTION)	GDP (CURRENT PRICES)	AGRICULTURE, FORESTRY AND FISHING VALUE ADDED SHARE IN GDP (2015 PRICES)	EMPLOYMENT IN AGRICULTURE, FORESTRY AND FISHING	SHARE OF EMPLOYMENT IN AGRICULTURE, FORESTRY AND FISHING IN TOTAL EMPLOYMENT
	MILLION, 2020	USD BILLION, 2020	PERCENT, 2020	THOUSAND, 2020	PERCENT, 2020
Jamaica	2.8	13.8	7.5	203.0	15.5
Japan	125.2	5 057.8	0.9	2 108.1	3.2
Jordan	10.9	43.7	4.8	69.7	3.2
Kazakhstan	19.0	171.1	5.0	1 329.7	15.0
Kenya	52.0	101.0	18.5	7 776.0	33.0
Kiribati	0.1	0.2	26.0		
Kuwait	4.4	105.9	0.6	45.8	2.0
Kyrgyzstan	6.4	7.7	14.5	380.6	16.6
Lao People's Democratic Republic	7.3	19.1	14.6	2 212.4	58.1
Latvia	1.9	33.7	3.6	58.7	6.8
Lebanon	5.7	63.5	4.4	70.7	3.8
Lesotho	2.3	2.3	4.6	210.4	30.0
Liberia	5.1	2.5	48.9	905.0	40.6
Libya	6.7	29.2	2.6	315.0	16.3
Lithuania	2.8	56.5	3.1	69.1	5.3
Luxembourg	0.6	73.4	0.2	3.6	1.1
Madagascar	28.2	13.0	26.3	10 410.4	73.9
Malawi	19.4	11.8	22.8	4 918.4	61.9
Malaysia	33.2	336.7	7.4	1 498.2	9.6
Maldives	0.5	3.7	7.2	24.2	10.5
Mali	21.2	17.3	35.2	4 760.7	67.7
Malta	0.5	14.9	0.5	2.0	0.9
Marshall Islands	0.0	0.2	14.7		

GROSS FIXED CAPITAL FORMATION (AGRICULTURE, FORESTRY AND FISHING) AS A SHARE OF VALUE ADDED	PESTICIDE USE, TOTAL	FERTILIZERS USE, TOTAL	INFLATION IN FOOD CONSUMER PRICES	CHANGE IN PRICES RECEIVED BY FARMERS	
PERCENT, 2021	THOUSAND TONNES, 2020	THOUSAND TONNES, 2020	PERCENT, 2021	PERCENT, 2021	
4.2	1.8	7.8	6.4	1.2	**Jamaica**
25.9	52.0	977.5	0.0	2.3	**Japan**
13.7	0.7	17.5	0.1	-1.0	**Jordan**
9.6	14.5	164.8	10.8	12.5	**Kazakhstan**
7.1	1.6	378.3	8.9	4.4	**Kenya**
6.7			0.9		**Kiribati**
20.0	0.0	9.0	9.5	29.3	**Kuwait**
7.9	0.7	29.2	18.0	0.2	**Kyrgyzstan**
13.2	0.2		3.0	0.3	**Lao People's Democratic Republic**
77.1	1.9	151.3	2.2	4.8	**Latvia**
10.7	1.8	33.7	309.9	3.2	**Lebanon**
10.8	0.0		9.7		**Lesotho**
5.9			0.4		**Liberia**
8.5	0.8	25.5	3.2		**Libya**
42.3	2.6	314.9	2.7	5.8	**Lithuania**
124.0	0.1	14.8	0.9	2.2	**Luxembourg**
5.0	0.8	31.7	7.4	3.1	**Madagascar**
11.1	2.4	117.3	11.3	9.0	**Malawi**
10.2	36.1	1 612.4	1.8	1.3	**Malaysia**
8.1	0.1	0.3	1.8	17.2	**Maldives**
6.5	0.0	238.7	5.0	3.6	**Mali**
38.8	0.1	1.1	2.3	-8.1	**Malta**
7.2					**Marshall Islands**

SELECTED INDICATORS – SOCIOECONOMIC (CONTINUED)

	POPULATION (PROJECTION)	GDP (CURRENT PRICES)	AGRICULTURE, FORESTRY AND FISHING VALUE ADDED SHARE IN GDP (2015 PRICES)	EMPLOYMENT IN AGRICULTURE, FORESTRY AND FISHING	SHARE OF EMPLOYMENT IN AGRICULTURE, FORESTRY AND FISHING IN TOTAL EMPLOYMENT
	MILLION, 2020	USD BILLION, 2020	PERCENT, 2020	THOUSAND, 2020	PERCENT, 2020
Mauritania	4.5	7.9	20.1	337.5	29.5
Mauritius	1.3	10.9	3.3	28.7	5.1
Mexico	126.0	1 073.4	3.6	6 756.9	12.3
Micronesia (Federated States of)	0.1	0.4	24.9		
Mongolia	3.3	13.1	15.3	304.1	24.3
Montenegro	0.6	4.8	8.3	16.9	7.4
Morocco	36.7	114.7	12.3	3 659.5	34.6
Mozambique	31.2	14.0	23.8	9 548.0	70.3
Myanmar	53.4	70.3	24.8	10 085.0	45.9
Namibia	2.5	10.7	8.1	165.1	22.1
Nauru	0.0	0.1	2.3		
Nepal	29.3	33.1	25.0	10 047.1	62.3
Netherlands	17.4	913.9	1.7	210.0	2.3
New Zealand	5.1	212.0	3.9	159.6	6.1
Nicaragua	6.8	12.6	19.4	814.8	28.7
Niger	24.3	13.7	36.9	6 491.0	70.7
Nigeria	208.3	429.9	23.4	20 478.0	35.2
North Macedonia	2.1	12.3	8.8	83.7	10.8
Norway	5.4	362.5	1.5	66.9	2.3
Oman	4.5	63.4	3.4	109.8	4.1
Pakistan	227.2	257.8	21.9	26 494.4	37.5
Palau	0.0	0.3	3.0		
Panama	4.3	52.9	3.3	273.3	15.7

GROSS FIXED CAPITAL FORMATION (AGRICULTURE, FORESTRY AND FISHING) AS A SHARE OF VALUE ADDED	PESTICIDE USE, TOTAL	FERTILIZERS USE, TOTAL	INFLATION IN FOOD CONSUMER PRICES	CHANGE IN PRICES RECEIVED BY FARMERS	
PERCENT, 2021	THOUSAND TONNES, 2020	THOUSAND TONNES, 2020	PERCENT, 2021	PERCENT, 2021	
7.1	0.0		5.8		**Mauritania**
10.0	0.5	11.3	4.4	-1.8	**Mauritius**
2.8	41.7	2 139.7	7.2	3.3	**Mexico**
8.5			1.2		**Micronesia (Federated States of)**
21.4	0.1	44.5	13.4	11.1	**Mongolia**
7.3	0.1	2.8	3.5		**Montenegro**
20.0	13.7	445.1	0.6	-0.6	**Morocco**
5.6	0.5	63.4	10.9	18.8	**Mozambique**
8.0	11.7	653.4	5.5	3.5	**Myanmar**
16.4	0.1	3.0	5.7	-7.4	**Namibia**
6.3		0.0			**Nauru**
7.5	0.7	215.5	4.3	6.3	**Nepal**
38.6	11.3	279.2	-0.2	-3.4	**Netherlands**
25.2	5.3	909.5	1.9	6.1	**New Zealand**
8.6	7.8	115.8	6.3	-2.3	**Nicaragua**
2.2	0.0	10.4	7.8	6.0	**Niger**
7.6		686.2	20.4	-3.2	**Nigeria**
5.4	0.1	34.5	3.1	3.1	**North Macedonia**
28.7	0.7	167.7	-2.0	0.7	**Norway**
0.9	1.7	24.6	0.8	-3.5	**Oman**
12.7	11.9	4 798.8	10.5	-7.4	**Pakistan**
7.1			3.1		**Palau**
10.3	2.1	43.8	1.6	0.0	**Panama**

SELECTED INDICATORS – SOCIOECONOMIC (CONTINUED)

	POPULATION (PROJECTION)	GDP (CURRENT PRICES)	AGRICULTURE, FORESTRY AND FISHING VALUE ADDED SHARE IN GDP (2015 PRICES)	EMPLOYMENT IN AGRICULTURE, FORESTRY AND FISHING	SHARE OF EMPLOYMENT IN AGRICULTURE, FORESTRY AND FISHING IN TOTAL EMPLOYMENT
	MILLION, 2020	USD BILLION, 2020	PERCENT, 2020	THOUSAND, 2020	PERCENT, 2020
Papua New Guinea	9.7	23.6	18.5	468.6	17.2
Paraguay	6.6	35.3	9.7	679.5	19.7
Peru	33.3	203.2	8.2	4 922.3	27.9
Philippines	112.2	361.5	9.9	10 380.9	24.3
Poland	38.4	596.6	2.2	1 482.1	8.4
Portugal	10.3	228.5	2.0	246.5	5.2
Qatar	2.8	146.4	0.3	26.2	1.2
Republic of Korea	51.8	1 637.9	1.7	1 460.2	5.3
Republic of Moldova	3.1	11.9	10.1	472.8	37.6
Romania	19.4	248.7	3.7	1 488.8	18.6
Russian Federation	145.6	1 483.5	4.1	4 023.2	5.8
Rwanda	13.1	10.3	23.2	3 573.6	54.7
Saint Kitts and Nevis	0.0	0.9	1.3		
Saint Lucia	0.2	1.6	2.6	9.0	10.5
Saint Vincent and the Grenadines	0.1	0.8	6.2	4.5	10.3
Samoa	0.2	0.8	8.9	11.7	24.1
San Marino	0.0	1.6	0.0		
Sao Tome and Principe	0.2	0.5	10.1	10.6	18.0
Saudi Arabia	36.0	700.1	2.7	403.7	2.7
Senegal	16.4	24.4	16.0	919.7	21.6
Serbia	7.4	53.3	6.5	490.8	13.9
Seychelles	0.1	1.1	2.2		

GROSS FIXED CAPITAL FORMATION (AGRICULTURE, FORESTRY AND FISHING) AS A SHARE OF VALUE ADDED	PESTICIDE USE, TOTAL	FERTILIZERS USE, TOTAL	INFLATION IN FOOD CONSUMER PRICES	CHANGE IN PRICES RECEIVED BY FARMERS	
PERCENT, 2021	THOUSAND TONNES, 2020	THOUSAND TONNES, 2020	PERCENT, 2021	PERCENT, 2021	
6.9	0.1	34.9	4.0		**Papua New Guinea**
9.0	20.2	584.3	9.1	10.1	**Paraguay**
10.4	10.6	556.8	6.0	-1.1	**Peru**
14.6	24.9	1 169.2	5.2	2.9	**Philippines**
39.8	24.2	1 729.0	3.0	3.5	**Poland**
33.3	9.7	180.8	0.7	4.9	**Portugal**
	0.1	5.0	2.6		**Qatar**
14.5	16.3	490.1	5.9	3.1	**Republic of Korea**
10.1	3.4	82.9	7.0	11.1	**Republic of Moldova**
28.2	5.4	738.5	3.2	4.7	**Romania**
16.5	90.5	3 081.1	9.6	8.1	**Russian Federation**
7.2	2.0	29.8	-4.1	2.2	**Rwanda**
12.5	0.0	0.0	0.1	2.2	**Saint Kitts and Nevis**
5.6	0.2	0.6	-0.1	-0.1	**Saint Lucia**
6.4			2.3	0.9	**Saint Vincent and the Grenadines**
	0.2	0.0	3.4	15.0	**Samoa**
					San Marino
10.7			9.5		**Sao Tome and Principe**
10.7	7.3	300.6	5.5	-14.4	**Saudi Arabia**
9.1	0.6	80.9	2.9	1.1	**Senegal**
15.3		417.5	4.4	5.4	**Serbia**
5.0	0.0	0.1	14.8	2.8	**Seychelles**

SELECTED INDICATORS – SOCIOECONOMIC (CONTINUED)

	POPULATION (PROJECTION)	GDP (CURRENT PRICES)	AGRICULTURE, FORESTRY AND FISHING VALUE ADDED SHARE IN GDP (2015 PRICES)	EMPLOYMENT IN AGRICULTURE, FORESTRY AND FISHING	SHARE OF EMPLOYMENT IN AGRICULTURE, FORESTRY AND FISHING IN TOTAL EMPLOYMENT
	MILLION, 2020	USD BILLION, 2020	PERCENT, 2020	THOUSAND, 2020	PERCENT, 2020
Sierra Leone	8.2	3.8	60.9	1 106.1	42.7
Singapore	5.9	340.0	0.0	11.1	0.3
Slovakia	5.5	105.2	2.0	64.1	2.5
Slovenia	2.1	53.6	2.1	39.6	4.1
Solomon Islands	0.7	1.5	37.3	135.8	38.0
Somalia	16.5	1.9	56.6	629.0	26.3
South Africa	58.8	302.1	2.4	3 207.4	21.3
South Sudan	10.6	15.9	5.7	2 582.3	62.1
Spain	47.4	1 281.5	3.1	796.9	4.1
Sri Lanka	21.7	80.7	7.4	1 948.4	25.7
Sudan	44.4	62.1	32.6	4 252.7	40.6
Suriname	0.6	4.1	8.3	16.7	7.9
Sweden	10.4	541.1	1.4	97.7	2.0
Switzerland	8.6	752.2	0.6	103.0	2.2
Syrian Arab Republic	20.8	15.6	25.1	608.3	12.5
Tajikistan	9.5	8.0	21.0	967.8	42.6
Thailand	71.5	501.8	8.7	12 162.0	31.6
Timor-Leste	1.3	1.9	14.2	224.3	41.6
Togo	8.4	7.1	23.6	861.8	30.9
Tonga	0.1	0.5	15.2	9.4	30.2
Trinidad and Tobago	1.5	21.4	1.1	18.5	3.0
Tunisia	12.2	39.2	11.5	480.8	13.9
Türkiye	84.1	720.1	6.7	4 837.4	17.1

GROSS FIXED CAPITAL FORMATION (AGRICULTURE, FORESTRY AND FISHING) AS A SHARE OF VALUE ADDED	PESTICIDE USE, TOTAL	FERTILIZERS USE, TOTAL	INFLATION IN FOOD CONSUMER PRICES	CHANGE IN PRICES RECEIVED BY FARMERS	
PERCENT, 2021	THOUSAND TONNES, 2020	THOUSAND TONNES, 2020	PERCENT, 2021	PERCENT, 2021	
3.3			17.0		**Sierra Leone**
54.0			1.6	-0.6	**Singapore**
42.1	1.8	174.7	1.8	-0.3	**Slovakia**
41.6	0.9	46.3	-0.2	1.0	**Slovenia**
6.7			-3.0		**Solomon Islands**
5.3			2.7		**Somalia**
20.1	26.9	761.9	6.1	4.4	**South Africa**
			13.5		**South Sudan**
21.3	43.3	1 945.5	1.8	1.2	**Spain**
15.3	2.6	408.5	11.1	7.8	**Sri Lanka**
	4.1	149.1	235.0		**Sudan**
9.6	0.4	7.9	61.0	12.4	**Suriname**
33.1	1.7	296.9	0.4	4.5	**Sweden**
55.6	1.9	70.0	-1.6	0.1	**Switzerland**
13.6	1.4	31.0	87.2	2.0	**Syrian Arab Republic**
12.1	0.3	81.2	10.5	2.9	**Tajikistan**
18.5	19.0	2 407.1	-0.1	6.9	**Thailand**
8.6	0.0		5.8	0.1	**Timor-Leste**
3.7	1.4	7.1	9.7	-0.9	**Togo**
11.3	0.0	0.2	7.2		**Tonga**
3.3	1.2	8.2	4.4	2.1	**Trinidad and Tobago**
12.9	3.5	146.7	6.3	4.9	**Tunisia**
14.6	53.7	2 930.9	23.9	15.3	**Türkiye**

SELECTED INDICATORS – SOCIOECONOMIC (CONTINUED)

	POPULATION (PROJECTION)	GDP (CURRENT PRICES)	AGRICULTURE, FORESTRY AND FISHING VALUE ADDED SHARE IN GDP (2015 PRICES)	EMPLOYMENT IN AGRICULTURE, FORESTRY AND FISHING	SHARE OF EMPLOYMENT IN AGRICULTURE, FORESTRY AND FISHING IN TOTAL EMPLOYMENT
	MILLION, 2020	USD BILLION, 2020	PERCENT, 2020	THOUSAND, 2020	PERCENT, 2020
Turkmenistan	6.3	42.8	11.3	411.8	22.3
Tuvalu	0.0	0.1	8.6		
Uganda	44.4	38.7	22.7	10 596.5	62.9
Ukraine	43.9	155.6	11.3	2 692.4	14.7
United Arab Emirates	9.3	358.9	0.9	107.0	1.7
United Kingdom of Great Britain and Northern Ireland	67.1	2 764.2	0.6	345.2	1.0
United Republic of Tanzania	61.7	66.5	25.0	18 170.5	64.3
United States of America	335.9	20 893.7	1.1	2 597.8	1.7
Uruguay	3.4	53.6	7.0	129.4	8.4
Uzbekistan	33.5	57.7	28.0	3 102.9	23.9
Vanuatu	0.3	0.9	21.3	62.9	48.0
Venezuela (Bolivarian Republic of)	28.5	106.4	7.0	1 307.0	13.1
Viet Nam	96.6	271.2	14.4	15 955.6	29.0
Yemen	32.3	28.0	18.6	1 677.9	28.1
Zambia	18.9	18.1	5.2	4 001.4	58.7
Zimbabwe	15.7	21.8	8.7	4 332.5	61.6

Source: FAO. 2022. FAOSTAT: Annual population. In: *FAO*. Rome. Cited October 2022. http://www.fao.org/faostat/en/#data/OA; FAO. 2022. FAOSTAT: Macro Indicators. In: *FAO*. Rome. Cited October 2022. http://www.fao.org/faostat/en/#data/MK; FAO. 2022. FAOSTAT: Employment Indicators: Agriculture. In: *FAO*. Rome. Cited October 2022. http://www.fao.org/faostat/en/#data/OEA; FAO. 2022. FAOSTAT: Capital Stock. In: *FAO*. Rome. Cited October 2022. http://www.fao.org/faostat/en/#data/CS;

GROSS FIXED CAPITAL FORMATION (AGRICULTURE, FORESTRY AND FISHING) AS A SHARE OF VALUE ADDED	PESTICIDE USE, TOTAL	FERTILIZERS USE, TOTAL	INFLATION IN FOOD CONSUMER PRICES	CHANGE IN PRICES RECEIVED BY FARMERS	
PERCENT, 2021	THOUSAND TONNES, 2020	THOUSAND TONNES, 2020	PERCENT, 2021	PERCENT, 2021	
15.1	9.7	325.1		6.6	**Turkmenistan**
14.3					**Tuvalu**
10.0	0.1	16.9	1.4		**Uganda**
13.2	24.6	2 488.7	10.8	5.8	**Ukraine**
16.1		31.5	0.0		**United Arab Emirates**
46.9	14.9	1 394.0	0.3	4.0	**United Kingdom of Great Britain and Northern Ireland**
8.3	0.0	222.0	3.7	5.3	**United Republic of Tanzania**
26.1	407.8	19 899.4	3.5	1.4	**United States of America**
14.3	16.4	411.1	6.8	9.5	**Uruguay**
8.5		1 023.9	14.3		**Uzbekistan**
12.4	0.1		5.8	0.6	**Vanuatu**
	3.9	324.6	1 460.4		**Venezuela (Bolivarian Republic of)**
15.5	19.2	3 174.0	0.7	0.5	**Viet Nam**
	0.1	12.3	4.8	8.7	**Yemen**
6.2	4.2	303.1	27.7	2.3	**Zambia**
12.2	2.2	132.8	132.7		**Zimbabwe**

FAO. 2022. FAOSTAT: Pesticides Use. In: *FAO*. Rome. Cited October 2022. http://www.fao.org/faostat/en/#data/RP;
FAO. 2022. FAOSTAT: Fertilizers by Nutrient. In: *FAO*. Rome. Cited October 2022. http://www.fao.org/faostat/en/#data/RFN;
FAO. 2022. FAOSTAT: Consumer Price Indices. In: *FAO*. Rome. Cited October 2022. https://www.fao.org/faostat/en/#data/CP;
FAO. 2022. FAOSTAT: Producer prices. In: *FAO*. Rome. Cited October 2022. http://www.fao.org/faostat/en/#data/PP

SELECTED INDICATORS – PRODUCTION AND TRADE

	AGRICULTURE PRODUCTION INDEX (2014–2016=100)	PRODUCTION OF CEREALS	PRODUCTION OF MEAT	PRODUCTION OF MILK	PRODUCTION OF RAW SUGAR
	2020	MILLION TONNES, 2020	THOUSAND TONNES, 2020	THOUSAND TONNES, 2020	THOUSAND TONNES, 2019
WORLD	107.4	2 996.1	337 179.9	886 861.8	179 461.7
AFRICA	112.1	208.2	19 875.3	51 082.6	11 870.8
AMERICAS	108.1	789.4	110 326.1	196 645.2	56 834.7
ASIA	108.2	1 444.4	135 040.6	374 075.8	77 397.0
EUROPE	102.8	526.4	65 119.6	234 368.3	28 631.9
OCEANIA	91.9	27.7	6 818.4	30 689.9	4 727.3
Afghanistan	118.3	6.0	312.8	2 137.1	1.1
Albania	105.5	0.7	79.5	1 052.2	2.9
Algeria	112.4	4.4	798.9	3 354.7	
Andorra					
Angola	108.0	2.4	342.3	219.7	100.0
Antigua and Barbuda	88.4	0.0	0.2	3.1	
Argentina	110.3	86.6	6 226.8	11 113.2	1 893.0
Armenia	85.7	0.2	107.4	654.3	8.8
Australia	86.3	26.6	4 796.9	8 797.0	4 516.7
Austria	101.7	5.6	853.7	3 852.3	340.0
Azerbaijan	127.5	3.2	346.0	2 192.5	53.4
Bahamas	100.6	0.0	6.9	2.8	
Bahrain	131.2	0.0	40.6	11.8	
Bangladesh	111.3	60.0	722.4	3 578.4	100.0
Barbados	101.3	0.0	19.1	5.3	7.9
Belarus	106.2	8.4	1 261.0	7 765.3	638.5
Belgium	101.5	2.6	1 797.7	4 494.0	
Belize	94.9	0.1	22.2	4.5	198.1
Benin	123.1	2.2	84.5	156.4	1.3

CAPTURE FISHERIES AND AQUACULTURE PRODUCTION	AQUACULTURE PRODUCTION	PRODUCTION OF ROUNDWOOD	VALUE OF FOOD IMPORTS	VALUE OF FOOD EXPORTS	
THOUSAND TONNES, 2020	THOUSAND TONNES, 2020	MILLION M³, 2020	USD BILLION, 2020	USD BILLION, 2020	
177 756.8	87 500.9	3 912.0	1 455.4	1 422.2	**WORLD**
12 044.2	2 250.2	791.6	82.2	51.7	**AFRICA**
21 903.2	4 375.2	1 070.9	267.2	388.4	**AMERICAS**
124 960.2	77 384.4	1 158.9	509.4	312.4	**ASIA**
17 095.7	3 262.6	803.7	576.0	614.7	**EUROPE**
1 752.5	228.5	86.8	20.5	55.0	**OCEANIA**
10.1	8.1	3.7	2.3	0.6	**Afghanistan**
16.9	9.3	0.9	0.8	0.2	**Albania**
86.9	5.4	8.9	8.5	0.5	**Algeria**
0.0		0.0			**Andorra**
379.4	2.1	6.2	1.8	0.1	**Angola**
3.2	0.0		0.1	0.0	**Antigua and Barbuda**
839.7	2.1	17.0	3.5	26.3	**Argentina**
18.9	18.4	1.5	0.6	0.5	**Armenia**
283.8	106.1	36.8	14.2	28.1	**Australia**
4.9	4.5	16.8	13.0	13.5	**Austria**
2.6	0.5	0.4	1.7	0.7	**Azerbaijan**
7.3	0.0	0.1	0.4	0.1	**Bahamas**
14.4	0.0	0.0	1.6	0.4	**Bahrain**
4 503.4	2 583.9	25.7	8.8	0.8	**Bangladesh**
1.5	0.0	0.0	0.3	0.1	**Barbados**
10.2	9.3	27.0	3.3	5.2	**Belarus**
20.2	0.2	5.2	34.9	39.7	**Belgium**
187.5	0.6	0.2	0.1	0.2	**Belize**
77.0	3.0	7.1	0.9	0.2	**Benin**

SELECTED INDICATORS – PRODUCTION AND TRADE (CONTINUED)

	AGRICULTURE PRODUCTION INDEX (2014–2016=100)	PRODUCTION OF CEREALS	PRODUCTION OF MEAT	PRODUCTION OF MILK	PRODUCTION OF RAW SUGAR
	2020	MILLION TONNES, 2020	THOUSAND TONNES, 2020	THOUSAND TONNES, 2020	THOUSAND TONNES, 2019
Bhutan	102.6	0.1	6.9	179.8	
Bolivia (Plurinational State of)	109.4	2.9	921.6	602.3	508.4
Bosnia and Herzegovina	114.7	1.9	89.0	645.0	
Botswana	102.4	0.2	65.7	280.1	
Brazil	113.3	125.6	29 125.8	36 806.8	27 732.0
Brunei Darussalam	105.9	0.0	29.1	0.2	
Bulgaria	89.7	8.6	176.8	1 005.5	
Burkina Faso	106.0	5.1	263.1	451.8	28.0
Burundi	136.4	0.5	36.5	119.3	24.0
Cabo Verde	80.5	0.0	5.0	9.7	
Cambodia	111.6	11.9	181.6	24.1	120.0
Cameroon	104.2	3.7	306.1	253.3	130.0
Canada	113.4	65.0	5 196.3	9 331.1	140.0
Central African Republic	105.2	0.1	186.7	81.5	11.6
Chad	119.4	2.9	818.7	395.9	21.0
Chile	107.9	2.8	1 588.0	2 283.5	201.0
China	104.1	617.5	77 142.7	39 219.5	11 810.8
Colombia	103.3	4.9	2 820.1	7 071.4	2 204.0
Comoros	104.6	0.0	2.3	13.4	
Congo	101.4	0.0	59.0	4.1	70.0
Costa Rica	98.2	0.2	294.9	1 201.5	442.2
Côte d'Ivoire	123.2	2.8	318.0	34.8	225.0
Croatia	86.1	3.7	215.9	612.0	

CAPTURE FISHERIES AND AQUACULTURE PRODUCTION	AQUACULTURE PRODUCTION	PRODUCTION OF ROUNDWOOD	VALUE OF FOOD IMPORTS	VALUE OF FOOD EXPORTS	
THOUSAND TONNES, 2020	THOUSAND TONNES, 2020	MILLION M³, 2020	USD BILLION, 2020	USD BILLION, 2020	
0.2	0.2	5.4	0.2	0.0	**Bhutan**
11.2	3.7	4.0	0.7	0.9	**Bolivia (Plurinational State of)**
4.1	3.8	4.3	1.6	0.5	**Bosnia and Herzegovina**
0.2	0.1	0.8	0.8	0.1	**Botswana**
1 338.8	629.5	266.3	9.9	73.1	**Brazil**
16.6	3.5	0.1	0.5	0.0	**Brunei Darussalam**
21.3	15.0	5.4	3.8	4.6	**Bulgaria**
29.7	0.6	15.7	0.4	0.3	**Burkina Faso**
21.0	1.5	6.6	0.1	0.1	**Burundi**
19.3	0.0	0.2	0.2	0.0	**Cabo Verde**
932.3	399.4	7.4	2.7	1.2	**Cambodia**
285.2	3.6	14.6	1.3	1.1	**Cameroon**
900.8	171.0	132.2	35.2	51.3	**Canada**
29.2	0.2	2.8	0.1	0.0	**Central African Republic**
107.1	0.1	9.0	0.2	0.1	**Chad**
3 259.4	1 485.9	59.5	6.3	16.8	**Chile**
63 854.1	49 900.8	338.7	189.6	75.6	**China**
256.7	179.4	9.0	5.5	6.2	**Colombia**
20.8		0.4	0.1	0.0	**Comoros**
71.5	0.9	4.0	0.6	0.0	**Congo**
33.7	16.3	4.5	1.9	4.6	**Costa Rica**
108.0	4.6	11.7	2.4	7.0	**Côte d'Ivoire**
93.3	21.7	5.2	3.1	2.5	**Croatia**

SELECTED INDICATORS – PRODUCTION AND TRADE (CONTINUED)

	AGRICULTURE PRODUCTION INDEX (2014–2016=100)	PRODUCTION OF CEREALS	PRODUCTION OF MEAT	PRODUCTION OF MILK	PRODUCTION OF RAW SUGAR
	2020	MILLION TONNES, 2020	THOUSAND TONNES, 2020	THOUSAND TONNES, 2020	THOUSAND TONNES, 2019
Cuba	74.3	0.5	223.1	459.1	1 211.1
Cyprus	105.9	0.1	78.4	347.4	
Czechia	97.3	8.1	461.3	3 282.4	625.0
Democratic People's Republic of Korea	94.7	4.7	333.7	84.2	
Democratic Republic of the Congo	110.5	3.6	243.0	8.2	90.0
Denmark	102.6	9.5	1 885.8	5 666.0	400.0
Djibouti	122.5	0.0	11.3	15.0	0.0
Dominica	100.0	0.0	1.4	7.1	
Dominican Republic	117.0	1.0	442.2	891.5	506.4
Ecuador	95.7	2.7	807.8	1 792.8	560.5
Egypt	99.6	22.3	2 197.3	5 089.5	2 600.0
El Salvador	103.8	1.0	175.7	407.5	823.0
Equatorial Guinea	101.0	0.0	0.6		
Eritrea	103.0	0.3	40.4	146.1	
Estonia	107.7	1.6	77.3	849.0	
Eswatini	102.7	0.1	26.3	40.2	788.4
Ethiopia	119.8	30.2	918.6	5 058.8	500.0
Fiji	114.8	0.0	32.7	12.4	175.6
Finland	98.6	3.4	409.0	2 406.5	
France	92.9	56.8	5 365.1	26 152.1	4 897.0
Gabon	103.0	0.0	39.7	14.0	26.0
Gambia	101.2	0.2	9.0	78.4	

CAPTURE FISHERIES AND AQUACULTURE PRODUCTION	AQUACULTURE PRODUCTION	PRODUCTION OF ROUNDWOOD	VALUE OF FOOD IMPORTS	VALUE OF FOOD EXPORTS	
THOUSAND TONNES, 2020	THOUSAND TONNES, 2020	MILLION M³, 2020	USD BILLION, 2020	USD BILLION, 2020	
46.4	26.2	1.8	1.7	0.3	**Cuba**
8.6	7.3	0.0	1.0	0.5	**Cyprus**
24.1	20.4	33.3	8.8	7.0	**Czechia**
279.3	77.3	7.8	0.5	0.1	**Democratic People's Republic of Korea**
221.3	3.6	92.4	0.7	0.1	**Democratic Republic of the Congo**
775.5	42.6	3.8	12.2	18.7	**Denmark**
2.3		0.4	1.1	0.1	**Djibouti**
0.9	0.0	0.0	0.0	0.0	**Dominica**
12.6	2.7	1.0	2.7	1.2	**Dominican Republic**
1 409.7	774.5	7.5	1.6	11.0	**Ecuador**
2 010.6	1 591.9	18.2	13.1	4.6	**Egypt**
61.5	8.5	4.8	2.1	1.0	**El Salvador**
6.1	0.0	1.7	0.3	0.0	**Equatorial Guinea**
5.6	0.0	1.0	0.1	0.0	**Eritrea**
75.0	1.1	10.6	1.5	1.5	**Estonia**
0.2	0.1	2.3	0.3	0.5	**Eswatini**
60.5	0.5	117.4	1.5	1.5	**Ethiopia**
42.6	0.2	0.8	0.4	0.4	**Fiji**
156.2	15.1	60.2	5.1	1.7	**Finland**
605.0	191.0	47.7	54.5	62.1	**France**
29.1	0.1	4.0	0.6	0.0	**Gabon**
51.0	0.0	0.9	0.2	0.0	**Gambia**

SELECTED INDICATORS – PRODUCTION AND TRADE (CONTINUED)

	AGRICULTURE PRODUCTION INDEX (2014–2016=100)	PRODUCTION OF CEREALS	PRODUCTION OF MEAT	PRODUCTION OF MILK	PRODUCTION OF RAW SUGAR
	2020	MILLION TONNES, 2020	THOUSAND TONNES, 2020	THOUSAND TONNES, 2020	THOUSAND TONNES, 2019
Georgia	115.7	0.4	69.4	582.2	
Germany	94.6	43.3	7 817.5	33 188.9	
Ghana	115.6	4.6	301.0	45.9	
Greece	99.6	3.1	437.9	1 990.3	9.0
Grenada	91.6	0.0	1.2	0.6	
Guatemala	107.8	2.0	531.3	515.2	2 962.6
Guinea	134.2	4.7	154.3	231.4	23.0
Guinea-Bissau	105.0	0.3	27.1	41.7	
Guyana	113.7	0.7	51.7	55.0	92.2
Haiti	78.4	0.4	106.3	104.0	14.2
Honduras	106.4	0.8	267.9	683.4	539.1
Hungary	93.6	15.6	984.6	2 019.0	117.3
Iceland	103.2	0.0	34.1	155.5	0.0
India	115.6	335.0	7 475.5	183 955.5	34 300.0
Indonesia	119.1	77.1	4 673.5	1 574.5	2 258.1
Iran (Islamic Republic of)	99.1	22.0	3 111.0	8 364.0	1 250.0
Iraq	148.4	8.9	254.7	404.2	0.3
Ireland	115.8	1.9	1 185.4	8 561.5	
Israel	101.4	0.2	839.1	1 605.6	
Italy	99.8	16.9	3 432.1	13 509.5	300.0
Jamaica	101.1	0.0	139.4	205.0	56.5
Japan	99.1	10.9	4 137.8	7 441.0	777.0
Jordan	99.1	0.1	254.1	427.9	
Kazakhstan	117.9	20.2	1 161.1	6 042.0	62.1

CAPTURE FISHERIES AND AQUACULTURE PRODUCTION	AQUACULTURE PRODUCTION	PRODUCTION OF ROUNDWOOD	VALUE OF FOOD IMPORTS	VALUE OF FOOD EXPORTS	
THOUSAND TONNES, 2020	THOUSAND TONNES, 2020	MILLION M³, 2020	USD BILLION, 2020	USD BILLION, 2020	
233.2	2.0	0.6	1.0	0.8	**Georgia**
244.4	32.3	84.1	87.5	70.0	**Germany**
420.4	64.0	52.4	2.7	3.0	**Ghana**
203.1	131.6	1.4	6.2	6.8	**Greece**
1.8	0.0		0.1	0.0	**Grenada**
44.9	33.7	22.3	2.8	5.8	**Guatemala**
310.7	1.1	13.1	0.9	0.1	**Guinea**
62.4	0.0	3.1	0.2	0.1	**Guinea-Bissau**
40.1	0.1	1.1	0.3	0.4	**Guyana**
17.9	1.6	2.4	1.4	0.1	**Haiti**
87.0	71.2	9.0	1.6	2.7	**Honduras**
23.8	18.4	5.6	5.5	8.5	**Hungary**
1 060.3	40.6	0.0	0.6	2.1	**Iceland**
14 140.7	8 636.0	350.7	18.9	32.3	**India**
12 151.6	5 226.6	122.0	14.2	33.2	**Indonesia**
1 282.4	480.5	0.4	6.9	3.0	**Iran (Islamic Republic of)**
57.5	22.7	0.2	9.6	0.1	**Iraq**
216.7	37.7	3.9	9.4	14.4	**Ireland**
16.7	14.7	0.0	6.1	1.8	**Israel**
263.3	122.8	15.8	41.4	46.8	**Italy**
13.8	0.9	0.7	0.8	0.3	**Jamaica**
3 751.2	599.5	30.3	57.2	7.2	**Japan**
2.6	2.1	0.4	3.7	1.0	**Jordan**
52.6	6.8	0.5	3.6	3.1	**Kazakhstan**

SELECTED INDICATORS – PRODUCTION AND TRADE (CONTINUED)

	AGRICULTURE PRODUCTION INDEX (2014–2016=100)	PRODUCTION OF CEREALS	PRODUCTION OF MEAT	PRODUCTION OF MILK	PRODUCTION OF RAW SUGAR
	2020	MILLION TONNES, 2020	THOUSAND TONNES, 2020	THOUSAND TONNES, 2020	THOUSAND TONNES, 2019
Kenya	110.0	4.9	592.9	5 515.7	440.9
Kiribati	94.4	0.0	2.0		
Kuwait	118.4	0.0	119.8	73.7	
Kyrgyzstan	110.3	1.9	238.5	1 707.2	99.7
Lao People's Democratic Republic	107.2	4.8	199.9	7.3	157.7
Latvia	110.9	3.5	89.5	990.1	
Lebanon	100.2	0.2	178.6	371.0	0.0
Lesotho	98.8	0.1	10.4	173.3	
Liberia	104.5	0.3	40.8	9.4	4.6
Libya	104.6	0.2	184.9	231.2	
Lithuania	107.6	6.5	222.1	1 491.7	
Luxembourg	113.8	0.1	22.9	450.5	
Madagascar	102.0	4.5	161.4	491.3	90.0
Malawi	129.3	4.0	450.8	211.2	252.0
Malaysia	99.7	2.4	1 866.5	50.7	1.2
Maldives	105.5	0.0	0.9		
Mali	125.2	10.4	182.3	996.0	114.0
Malta	74.2	0.0	10.0	44.9	
Marshall Islands	96.1	0.0			
Mauritania	110.5	0.5	120.3	364.3	
Mauritius	80.8	0.0	50.4	1.8	331.1
Mexico	112.8	36.4	7 536.8	12 783.7	6 710.1
Micronesia (Federated States of)	99.7	0.0	1.5		

CAPTURE FISHERIES AND AQUACULTURE PRODUCTION	AQUACULTURE PRODUCTION	PRODUCTION OF ROUNDWOOD	VALUE OF FOOD IMPORTS	VALUE OF FOOD EXPORTS	
THOUSAND TONNES, 2020	THOUSAND TONNES, 2020	MILLION M³, 2020	USD BILLION, 2020	USD BILLION, 2020	
142.8	20.0	25.9	2.6	2.7	Kenya
212.8	0.0	0.0	0.0	0.2	Kiribati
3.5	0.5	0.0	5.2	0.4	Kuwait
2.6	2.6	0.0	0.5	0.2	Kyrgyzstan
200.0	130.0	7.1	1.0	0.8	Lao People's Democratic Republic
104.2	0.7	15.3	3.1	3.4	Latvia
3.7	0.8	0.0	2.1	0.7	Lebanon
2.7	2.6	2.2	0.3	0.0	Lesotho
31.9	0.3	10.2	0.4	0.1	Liberia
31.6	0.0	1.2	2.9	0.0	Libya
95.6	4.5	6.4	3.9	5.1	Lithuania
0.0		0.3	2.4	1.2	Luxembourg
115.8	5.5	15.4	0.6	0.9	Madagascar
180.5	9.4	7.5	0.3	0.3	Malawi
1 606.9	218.0	17.2	14.8	20.7	Malaysia
148.6		0.0	0.4	0.2	Maldives
125.4	7.7	6.7	0.6	0.2	Mali
21.9	19.8	0.0	0.6	0.3	Malta
88.2	0.0				Marshall Islands
678.4		2.3	0.8	0.6	Mauritania
29.3	3.3	0.0	0.9	0.6	Mauritius
1 779.7	278.7	46.1	19.8	35.1	Mexico
193.6	0.0	0.0			Micronesia (Federated States of)

SELECTED INDICATORS – PRODUCTION AND TRADE (CONTINUED)

	AGRICULTURE PRODUCTION INDEX (2014–2016=100)	PRODUCTION OF CEREALS	PRODUCTION OF MEAT	PRODUCTION OF MILK	PRODUCTION OF RAW SUGAR
	2020	MILLION TONNES, 2020	THOUSAND TONNES, 2020	THOUSAND TONNES, 2020	THOUSAND TONNES, 2019
Mongolia	171.5	0.4	625.1	1 082.4	
Montenegro	102.5	0.0	14.1	175.8	
Morocco	100.6	3.3	1 447.1	2 591.7	625.0
Mozambique	132.9	1.9	229.1	557.9	440.0
Myanmar	102.8	27.6	3 465.4	2 465.9	26.8
Namibia	103.2	0.2	72.8	112.0	
Nauru	101.3	0.0	0.1		
Nepal	114.6	10.9	412.4	2 455.3	180.0
Netherlands	102.0	1.4	3 108.8	14 932.0	
New Zealand	102.8	1.0	1 454.3	21 871.3	
Nicaragua	133.5	0.9	307.7	1 399.8	757.5
Niger	129.2	5.9	196.3	1 511.0	30.5
Nigeria	106.4	28.7	1 450.2	524.7	35.0
North Macedonia	102.1	0.6	23.5	447.7	0.3
Norway	100.3	1.2	361.3	1 565.3	0.0
Oman	152.6	0.2	88.0	308.9	
Pakistan	114.0	42.5	4 737.1	60 770.0	4 881.2
Palau					
Panama	106.6	0.5	324.1	213.8	161.6
Papua New Guinea	101.7	0.0	507.0	0.2	35.0
Paraguay	116.8	8.4	659.7	541.0	142.8
Peru	119.2	5.4	2 156.8	2 158.5	1 196.0
Philippines	100.6	27.4	3 141.6	17.2	2 036.6
Poland	110.8	34.9	5 216.3	14 830.9	

CAPTURE FISHERIES AND AQUACULTURE PRODUCTION	AQUACULTURE PRODUCTION	PRODUCTION OF ROUNDWOOD	VALUE OF FOOD IMPORTS	VALUE OF FOOD EXPORTS	
THOUSAND TONNES, 2020	THOUSAND TONNES, 2020	MILLION M³, 2020	USD BILLION, 2020	USD BILLION, 2020	
0.0		0.9	0.7	0.1	**Mongolia**
1.7	0.9	1.1	0.5	0.0	**Montenegro**
1 376.7	1.4	6.7	5.7	6.0	**Morocco**
403.1	3.2	18.7	1.3	0.4	**Mozambique**
2 998.6	1 145.0	42.6	2.2	4.6	**Myanmar**
329.9	0.3	2.0	0.8	0.9	**Namibia**
92.4	0.0				**Nauru**
97.8	76.8	13.0	1.9	0.4	**Nepal**
345.0	39.9	3.1	63.1	84.5	**Netherlands**
482.5	118.6	36.0	4.0	24.9	**New Zealand**
79.1	29.4	6.3	0.9	2.2	**Nicaragua**
46.6	0.6	12.8	0.9	0.2	**Niger**
1 044.8	261.7	76.9	7.1	1.3	**Nigeria**
2.1	1.6	0.7	0.8	0.4	**North Macedonia**
3 941.0	1 490.1	11.8	6.5	11.5	**Norway**
794.7	1.3	0.1	3.8	0.9	**Oman**
655.2	162.5	33.6	6.8	4.0	**Pakistan**
0.8	0.0				**Palau**
191.6	3.9	1.4	2.2	0.4	**Panama**
219.4	1.8	9.6	0.8	1.0	**Papua New Guinea**
31.1	14.1	11.7	0.7	4.8	**Paraguay**
5 770.4	143.8	7.3	4.3	8.4	**Peru**
2 766.1	854.2	15.2	10.2	5.4	**Philippines**
254.1	47.7	40.6	20.0	30.3	**Poland**

SELECTED INDICATORS – PRODUCTION AND TRADE (CONTINUED)

	AGRICULTURE PRODUCTION INDEX (2014–2016=100)	PRODUCTION OF CEREALS	PRODUCTION OF MEAT	PRODUCTION OF MILK	PRODUCTION OF RAW SUGAR
	2020	MILLION TONNES, 2020	THOUSAND TONNES, 2020	THOUSAND TONNES, 2020	THOUSAND TONNES, 2019
Portugal	107.0	1.1	846.9	2 099.1	
Qatar	146.3	0.0	42.5	51.6	
Republic of Korea	98.4	4.9	2 743.4	1 806.0	
Republic of Moldova	81.8	1.5	129.7	329.7	89.9
Romania	90.6	19.4	530.3	4 362.5	
Russian Federation	112.0	130.0	11 222.0	32 219.2	7 309.7
Rwanda	111.9	0.8	109.2	253.4	10.4
Saint Kitts and Nevis	91.1	0.0	0.1		
Saint Lucia	86.4	0.0	2.8	1.0	
Saint Vincent and the Grenadines	105.7	0.0	1.1	1.0	
Samoa	94.5	0.0	2.9	1.8	
San Marino					
Sao Tome and Principe	103.3	0.0	1.3	0.6	
Saudi Arabia	150.8	1.2	1 188.9	2 911.2	
Senegal	180.2	3.6	287.7	247.2	160.0
Serbia	110.9	11.5	517.0	1 583.7	450.0
Seychelles	98.1	0.0	1.0	0.1	
Sierra Leone	108.3	1.2	46.9	150.6	8.0
Singapore	160.3	0.0	127.6		
Slovakia	101.1	4.6	82.2	929.5	215.0
Slovenia	104.8	0.7	140.2	633.2	
Solomon Islands	104.9	0.0	3.6	3.1	

CAPTURE FISHERIES AND AQUACULTURE PRODUCTION	AQUACULTURE PRODUCTION	PRODUCTION OF ROUNDWOOD	VALUE OF FOOD IMPORTS	VALUE OF FOOD EXPORTS	
THOUSAND TONNES, 2020	THOUSAND TONNES, 2020	MILLION M³, 2020	USD BILLION, 2020	USD BILLION, 2020	
173.3	14.6	13.3	11.0	7.2	**Portugal**
15.1	0.0	0.0	2.9	0.0	**Qatar**
1 934.3	566.4	4.3	27.7	6.9	**Republic of Korea**
12.7	12.6	1.3	0.7	1.0	**Republic of Moldova**
20.2	12.2	15.5	8.6	6.0	**Romania**
5 342.5	270.4	217.0	25.8	26.5	**Russian Federation**
37.0	7.1	6.2	0.4	0.3	**Rwanda**
0.6	0.0		0.0	0.0	**Saint Kitts and Nevis**
1.4	0.0	0.0	0.1	0.0	**Saint Lucia**
2.1		0.0	0.1	0.0	**Saint Vincent and the Grenadines**
10.0	0.0	0.1	0.1	0.0	**Samoa**
0.0					**San Marino**
5.6		0.1	0.0	0.0	**Sao Tome and Principe**
161.8	99.9	0.3	19.2	3.5	**Saudi Arabia**
452.8	1.1	6.4	1.9	1.2	**Senegal**
7.9	6.0	8.2	1.9	3.4	**Serbia**
132.4	0.0	0.0	0.3	0.5	**Seychelles**
200.7	0.1	6.4	0.4	0.1	**Sierra Leone**
6.2	4.8	0.0	11.4	11.7	**Singapore**
4.0	2.3	7.4	4.8	3.3	**Slovakia**
2.0	1.7	3.9	2.7	2.0	**Slovenia**
40.9	0.0	3.3	0.1	0.1	**Solomon Islands**

SELECTED INDICATORS – PRODUCTION AND TRADE (CONTINUED)

	AGRICULTURE PRODUCTION INDEX (2014–2016=100)	PRODUCTION OF CEREALS	PRODUCTION OF MEAT	PRODUCTION OF MILK	PRODUCTION OF RAW SUGAR
	2020	MILLION TONNES, 2020	THOUSAND TONNES, 2020	THOUSAND TONNES, 2020	THOUSAND TONNES, 2019
Somalia	99.6	0.2	194.2	2 209.2	22.0
South Africa	111.2	18.2	3 447.2	3 821.5	2 295.0
South Sudan	102.5	0.9	251.6	3 057.3	
Spain	117.5	27.3	7 438.5	8 686.2	490.0
Sri Lanka	119.0	5.4	253.0	507.6	65.0
Sudan	116.1	3.8	1 003.7	4 655.2	542.0
Suriname	99.8	0.3	14.4	3.7	6.3
Sweden	101.1	6.0	567.1	2 772.7	
Switzerland	98.3	1.0	482.1	3 794.6	237.2
Syrian Arab Republic	122.2	5.3	352.7	2 153.7	3.3
Tajikistan	134.7	1.3	232.6	1 031.9	
Thailand	96.7	35.5	2 871.3	1 200.0	14 866.8
Timor-Leste	100.4	0.1	35.3	6.1	
Togo	110.7	1.4	90.4	11.5	
Tonga	94.1	0.0	2.4	0.3	
Trinidad and Tobago	104.3	0.0	66.5	2.2	
Tunisia	122.3	1.6	346.0	1 437.1	15.0
Türkiye	114.5	37.2	3 275.4	21 839.4	2 494.1
Turkmenistan	95.9	1.5	302.4	1 792.9	30.0
Tuvalu	101.8	0.0	0.2		
Uganda	117.1	3.4	439.3	1 766.4	514.0
Ukraine	99.3	64.3	2 486.9	9 263.9	1 490.0
United Arab Emirates	108.3	0.0	173.2	164.9	

CAPTURE FISHERIES AND AQUACULTURE PRODUCTION	AQUACULTURE PRODUCTION	PRODUCTION OF ROUNDWOOD	VALUE OF FOOD IMPORTS	VALUE OF FOOD EXPORTS	
THOUSAND TONNES, 2020	THOUSAND TONNES, 2020	MILLION M³, 2020	USD BILLION, 2020	USD BILLION, 2020	
30.0		16.5	1.8	0.3	Somalia
601.9	6.0	28.7	5.3	9.3	South Africa
30.0	0.0	4.8	0.4	0.0	South Sudan
1 078.5	276.6	18.3	34.2	56.6	Spain
428.5	41.7	5.1	2.2	2.6	Sri Lanka
47.5	9.9	16.7	2.3	1.3	Sudan
29.8	0.0	1.2	0.2	0.1	Suriname
191.7	12.1	76.1	15.9	9.7	Sweden
3.5	2.0	4.7	11.9	8.8	Switzerland
6.1	2.3	0.1	1.5	0.6	Syrian Arab Republic
2.7	0.8	3.7	0.7	0.0	Tajikistan
2 617.8	962.5	33.0	13.4	32.0	Thailand
8.4	0.1	0.1	0.2	0.0	Timor-Leste
18.8	0.7	4.6	0.4	0.2	Togo
1.1	0.0	0.0	0.1	0.0	Tonga
12.9	0.0	0.2	0.8	0.2	Trinidad and Tobago
126.6	23.4	3.9	2.1	1.5	Tunisia
785.8	421.4	28.7	12.1	18.8	Türkiye
14.8	0.2	0.0	0.4	0.1	Turkmenistan
11.3	0.0		0.0		Tuvalu
690.2	123.9	50.0	0.9	1.5	Uganda
87.2	18.6	16.8	5.4	19.9	Ukraine
69.5	3.0	0.0	14.8	8.7	United Arab Emirates

SELECTED INDICATORS – PRODUCTION AND TRADE (CONTINUED)

	AGRICULTURE PRODUCTION INDEX (2014–2016=100)	PRODUCTION OF CEREALS	PRODUCTION OF MEAT	PRODUCTION OF MILK	PRODUCTION OF RAW SUGAR
	2020	MILLION TONNES, 2020	THOUSAND TONNES, 2020	THOUSAND TONNES, 2020	THOUSAND TONNES, 2019
United Kingdom of Great Britain and Northern Ireland	96.9	19.0	4 173.6	15 558.0	1 188.8
United Republic of Tanzania	116.5	12.5	685.7	3 221.2	370.0
United States of America	104.3	434.9	48 710.8	101 277.0	7 374.0
Uruguay	91.8	3.5	581.3	2 205.0	25.0
Uzbekistan	104.6	7.1	1 214.8	10 930.1	
Vanuatu	89.8	0.0	5.9	2.4	
Venezuela (Bolivarian Republic of)	85.6	2.0	952.4	2 171.6	369.2
Viet Nam	107.5	47.3	5 388.9	1 076.2	1 812.8
Yemen	105.3	0.4	417.4	327.1	
Zambia	116.4	3.7	332.2	389.1	450.0
Zimbabwe	113.3	1.6	193.7	426.8	482.9

Source: FAO. 2022. FAOSTAT: Production Indices. In: *FAO*. Rome. Cited October 2022. http://www.fao.org/faostat/en/#data/QI; FAO. 2022. FAOSTAT: Production: Crops and livestock products. In: *FAO*. Rome. Cited October 2022. https://www.fao.org/faostat/en/#data/QCL;
FAO. 2022. Fisheries and Aquaculture: Global production by production source Quantity (1950 - 2020). In: *FAO*. Rome. Cited October 2022. https://www.fao.org/fishery/statistics-query/en/global_production/global_production_quantity;
FAO. 2022. FAOSTAT: Forestry Production and Trade. In: *FAO*. Rome. Cited October 2022. https://www.fao.org/faostat/en/#data/FO;

CAPTURE FISHERIES AND AQUACULTURE PRODUCTION	AQUACULTURE PRODUCTION	PRODUCTION OF ROUNDWOOD	VALUE OF FOOD IMPORTS	VALUE OF FOOD EXPORTS	
THOUSAND TONNES, 2020	THOUSAND TONNES, 2020	MILLION M³, 2020	USD BILLION, 2020	USD BILLION, 2020	
847.0	221.0	10.5	58.9	26.7	**United Kingdom of Great Britain and Northern Ireland**
485.8	17.5	28.2	0.8	1.4	**United Republic of Tanzania**
4 694.3	448.2	429.7	155.2	128.9	**United States of America**
63.0	0.1	18.0	1.1	4.5	**Uruguay**
144.1	98.0	0.0	2.1	1.3	**Uzbekistan**
51.5	0.0	0.1	0.1	0.1	**Vanuatu**
307.5	49.1	5.7	2.6	0.2	**Venezuela (Bolivarian Republic of)**
8 022.7	4 600.8	57.3	16.1	24.4	**Viet Nam**
131.3	0.0	0.6	4.1	0.3	**Yemen**
152.5	45.7	25.7	0.4	0.4	**Zambia**
34.2	15.4	10.0	0.9	0.2	**Zimbabwe**

FAO. 2022. FAOSTAT: Trade: Crops and livestock products. In: *FAO*. Rome. Cited October 2022. https://www.fao.org/faostat/en/#data/TCL and FAO. 2022. Fisheries and Aquaculture: Global fish trade - All partners aggregated Value (1976 - 2020). In: *FAO*. Rome. Cited October 2022. https://www.fao.org/fishery/statistics-query/en/trade/trade_value

SELECTED INDICATORS – FOOD SECURITY AND NUTRITION

	PREVALENCE OF UNDERNOURISHMENT	NUMBER OF UNDERNOURISHED	PREVALENCE OF SEVERE FOOD INSECURITY	PREVALENCE OF MODERATE OR SEVERE FOOD INSECURITY	AVERAGE DIETARY ENERGY SUPPLY
	PERCENT, 2019–2021	MILLION, 2019–2021	PERCENT, 2019–2021	PERCENT, 2019–2021	KCAL/CAP/DAY, 2019–2021
WORLD	9.0	702.7	10.7	28.1	2 963
AFRICA	19.1	256.1	22.0	55.5	2 589
ASIA	8.3	387.5	9.5	23.9	2 922
LATIN AMERICA AND THE CARIBBEAN	7.7	50.7	12.3	37.3	3 035
NORTHERN AMERICA AND EUROPE	<2.5		1.2	7.6	3 537
OCEANIA	5.6	2.4	3.7	12.9	3 112
Afghanistan	29.8	11.6	22.5	70.0	2 239
Albania	3.9	0.1	7.7	30.9	3 378
Algeria	<2.5		6.2	19.0	3 479
Andorra					
Angola	20.8	6.8	30.4	77.7	2 401
Antigua and Barbuda			7.1	33.0	
Argentina	3.7	1.7	13.0	37.0	3 314
Armenia	3.5	0.1	1.0	10.7	2 996
Australia	<2.5		3.6	11.9	3 424
Austria	<2.5		1.3	3.3	3 672
Azerbaijan	<2.5		<0.5	9.5	3 145
Bahamas			3.4	17.2	
Bahrain					
Bangladesh	11.4	18.8	10.7	31.7	2 574
Barbados	3.4	<0.1	7.4	31.1	3 003
Belarus	<2.5				3 279
Belgium	<2.5		1.3	4.8	3 784

AVERAGE DIETARY ENERGY SUPPLY ADEQUACY	SHARE OF CEREALS/ROOTS/ TUBERS IN DIETARY ENERGY SUPPLY	CEREAL IMPORT DEPENDENCY RATIO	PREVALENCE OF STUNTING, CHILDREN UNDER 5	PREVALENCE OF OBESITY, ADULTS 18 YEARS AND OLDER	
PERCENT, 2019–2021	PERCENT, 2017–2019	PERCENT, 2017–2019	PERCENT, 2020	PERCENT, 2016	
124	50	-1.3	22.0	13.1	**WORLD**
115	62	29.6	30.7	12.8	**AFRICA**
123	53	8.3	21.8	7.3	**AMERICAS**
126	38	-7.2	11.3	24.2	**ASIA**
141	31	-27.4		26.9	**EUROPE**
126	31	-93.3		28.1	**OCEANIA**
104	70	41.4	35.1	5.5	**Afghanistan**
136	35	38.2	9.6	21.7	**Albania**
153	50	70.1	9.3	27.4	**Algeria**
				25.6	**Andorra**
113		41.5	37.7	8.2	**Angola**
		94.7		18.9	**Antigua and Barbuda**
137		-128.5	7.8	28.3	**Argentina**
123	40	66.0	9.1	20.2	**Armenia**
138	24	-134.5	2.1	29.0	**Australia**
145	27	16.5		20.1	**Austria**
130	58	30.2	16.3	19.9	**Azerbaijan**
		100.0		31.6	**Bahamas**
			5.1	29.8	**Bahrain**
112	78	12.3	30.2	3.6	**Bangladesh**
121	32	100.0	6.6	23.1	**Barbados**
134	36	9.3	3.9	24.5	**Belarus**
149	27	66.6	2.3	22.1	**Belgium**

SELECTED INDICATORS – FOOD SECURITY AND NUTRITION (CONTINUED)

	PREVALENCE OF UNDERNOURISHMENT	NUMBER OF UNDERNOURISHED	PREVALENCE OF SEVERE FOOD INSECURITY	PREVALENCE OF MODERATE OR SEVERE FOOD INSECURITY	AVERAGE DIETARY ENERGY SUPPLY
	PERCENT, 2019–2021	MILLION, 2019–2021	PERCENT, 2019–2021	PERCENT, 2019–2021	KCAL/CAP/DAY, 2019–2021
Belize	7.4	<0.1	6.0	42.3	2 703
Benin	7.4	0.9	13.8	67.9	2 821
Bhutan					
Bolivia (Plurinational State of)	13.9	1.6			2 486
Bosnia and Herzegovina	<2.5		2.8	12.6	3 318
Botswana	21.9	0.5	25.4	55.6	2 605
Brazil	4.1	8.6	7.3	28.9	3 235
Brunei Darussalam	5.9	<0.1			2 816
Bulgaria	3.0	0.2	2.9	15.5	2 875
Burkina Faso	18.0	3.8	18.5	52.6	2 671
Burundi					
Cabo Verde	17.7	<0.1	6.4	35.4	2 516
Cambodia	6.3	1.0	15.1	50.0	2 757
Cameroon	6.7	1.8	26.7	55.8	2 791
Canada	<2.5		1.0	6.5	3 524
Central African Republic	52.2	2.5	61.8	81.3	1 829
Chad	32.7	5.4			2 258
Chile	2.6	0.5	3.8	17.4	3 081
China	<2.5				3 336
Colombia	8.2	4.2			2 968
Comoros	20.4	0.2	27.4	79.7	
Congo	31.6	1.7	55.5	88.7	2 207
Costa Rica	3.4	0.2	2.8	15.9	3 000

AVERAGE DIETARY ENERGY SUPPLY ADEQUACY	SHARE OF CEREALS/ROOTS/ TUBERS IN DIETARY ENERGY SUPPLY	CEREAL IMPORT DEPENDENCY RATIO	PREVALENCE OF STUNTING, CHILDREN UNDER 5	PREVALENCE OF OBESITY, ADULTS 18 YEARS AND OLDER	
PERCENT, 2019–2021	PERCENT, 2017–2019	PERCENT, 2017–2019	PERCENT, 2020	PERCENT, 2016	
118	37	18.1	13.3	24.1	**Belize**
128	68	46.0	31.3	9.6	**Benin**
			22.4	6.4	**Bhutan**
110	48	18.6	12.7	20.2	**Bolivia (Plurinational State of)**
132		30.6	9.1	17.9	**Bosnia and Herzegovina**
112	55	90.7	22.8	18.9	**Botswana**
132	31	-26.0	6.1	22.1	**Brazil**
117			12.7	14.1	**Brunei Darussalam**
116	39	-157.3	6.4	25.0	**Bulgaria**
121	64	9.2	25.5	5.6	**Burkina Faso**
		25.7	57.6	5.4	**Burundi**
105	52	100.0	9.7	11.8	**Cabo Verde**
122	68	-6.6	29.9	3.9	**Cambodia**
124	56	30.7	27.2	11.4	**Cameroon**
141	28	-84.1		29.4	**Canada**
84	52		40.1	7.5	**Central African Republic**
103	62	4.0	35.0	6.1	**Chad**
126	43	46.9	1.6	28.0	**Chile**
136	49	3.8	4.7	6.2	**China**
126	34	62.9	11.5	22.3	**Colombia**
103	57		22.6	7.8	**Comoros**
99	61	94.8	18.0	9.6	**Congo**
123	31	86.9	8.6	25.7	**Costa Rica**

SELECTED INDICATORS – FOOD SECURITY AND NUTRITION (CONTINUED)

	PREVALENCE OF UNDERNOURISHMENT	NUMBER OF UNDERNOURISHED	PREVALENCE OF SEVERE FOOD INSECURITY	PREVALENCE OF MODERATE OR SEVERE FOOD INSECURITY	AVERAGE DIETARY ENERGY SUPPLY
	PERCENT, 2019–2021	MILLION, 2019–2021	PERCENT, 2019–2021	PERCENT, 2019–2021	KCAL/CAP/DAY, 2019–2021
Côte d'Ivoire	4.4	1.2	9.4	42.8	2 958
Croatia	<2.5		1.6	11.4	3 146
Cuba	<2.5				3 202
Cyprus	<2.5				3 026
Czechia	<2.5		1.6	5.8	3 280
Democratic People's Republic of Korea	41.6	10.7			2 084
Democratic Republic of the Congo	39.8	35.6	39.2	72.3	2 064
Denmark	<2.5		1.4	5.5	3 348
Djibouti	13.5	0.1	16.5	49.2	2 776
Dominica	6.9	<0.1			2 877
Dominican Republic	6.7	0.7			2 997
Ecuador	15.4	2.7	12.8	36.8	2 502
Egypt	5.1	5.2	7.1	27.3	3 306
El Salvador	7.7	0.5	14.7	46.5	2 735
Equatorial Guinea					
Eritrea					
Estonia	<2.5		0.8	7.9	3 169
Eswatini	11.0	0.1	18.3	67.0	2 564
Ethiopia	24.9	28.6	19.6	56.2	2 404
Fiji	5.7	<0.1	4.2	19.3	2 874
Finland	<2.5		2.4	8.8	3 314
France	<2.5		1.0	5.9	3 515

AVERAGE DIETARY ENERGY SUPPLY ADEQUACY	SHARE OF CEREALS/ROOTS/ TUBERS IN DIETARY ENERGY SUPPLY	CEREAL IMPORT DEPENDENCY RATIO	PREVALENCE OF STUNTING, CHILDREN UNDER 5	PREVALENCE OF OBESITY, ADULTS 18 YEARS AND OLDER	
PERCENT, 2019–2021	PERCENT, 2017–2019	PERCENT, 2017–2019	PERCENT, 2020	PERCENT, 2016	
133	72	42.1	17.8	10.3	**Côte d'Ivoire**
127		-31.1		24.4	**Croatia**
131	46		7.0	24.6	**Cuba**
120	40	98.1		21.8	**Cyprus**
130	27	-55.7	2.5	26.0	**Czechia**
87	68	18.3	18.2	6.8	**Democratic People's Republic of Korea**
97	73	19.1	40.8	6.7	**Democratic Republic of the Congo**
131	28	-9.3		19.7	**Denmark**
115		97.9	34.0	13.5	**Djibouti**
152	33	100.0		27.9	**Dominica**
125	28	64.8	5.9	27.6	**Dominican Republic**
109	43	36.3	23.1	19.9	**Ecuador**
143	66	47.8	22.3	32.0	**Egypt**
120		51.9	11.2	24.6	**El Salvador**
			19.7	8.0	**Equatorial Guinea**
			49.1	5.0	**Eritrea**
125		-130.0	1.2	21.2	**Estonia**
112	55	61.0	22.6	16.5	**Eswatini**
108	75	7.3	35.3	4.5	**Ethiopia**
121		100.0	7.5	30.2	**Fiji**
130		-12.1		22.2	**Finland**
141		-71.2		21.6	**France**

SELECTED INDICATORS – FOOD SECURITY AND NUTRITION (CONTINUED)

	PREVALENCE OF UNDERNOURISHMENT	NUMBER OF UNDERNOURISHED	PREVALENCE OF SEVERE FOOD INSECURITY	PREVALENCE OF MODERATE OR SEVERE FOOD INSECURITY	AVERAGE DIETARY ENERGY SUPPLY
	PERCENT, 2019–2021	MILLION, 2019–2021	PERCENT, 2019–2021	PERCENT, 2019–2021	KCAL/CAP/DAY, 2019–2021
Gabon	17.2	0.4			2 675
Gambia	21.6	0.5	27.1	58.0	2 336
Georgia	7.6	0.3	9.5	38.8	2 901
Germany	<2.5		1.1	3.5	3 549
Ghana	4.1	1.3	5.6	36.6	3 261
Greece	<2.5		1.6	6.8	3 412
Grenada			7.5	22.3	
Guatemala	16.0	2.9	20.7	55.9	2 593
Guinea			48.9	73.3	
Guinea-Bissau	31.7	0.6	29.2	75.0	
Guyana	4.9	<0.1			2 933
Haiti	47.2	5.4	45.2	82.5	2 087
Honduras	15.3	1.5	17.9	49.9	2 686
Hungary	<2.5		2.1	10.6	3 399
Iceland	<2.5		1.3	6.3	3 642
India	16.3	224.3			2 594
Indonesia	6.5	17.7	0.7	6.0	2 909
Iran (Islamic Republic of)	4.1	3.4	7.7	42.4	3 075
Iraq	15.9	6.4			2 702
Ireland	<2.5		3.2	6.5	3 769
Israel	<2.5		2.0	14.2	3 532
Italy	<2.5		1.9	6.3	3 509
Jamaica	6.9	0.2	23.1	50.3	2 831
Japan	3.2	4.0	0.9	3.8	2 640

AVERAGE DIETARY ENERGY SUPPLY ADEQUACY	SHARE OF CEREALS/ROOTS/ TUBERS IN DIETARY ENERGY SUPPLY	CEREAL IMPORT DEPENDENCY RATIO	PREVALENCE OF STUNTING, CHILDREN UNDER 5	PREVALENCE OF OBESITY, ADULTS 18 YEARS AND OLDER	
PERCENT, 2019–2021	PERCENT, 2017–2019	PERCENT, 2017–2019	PERCENT, 2020	PERCENT, 2016	
118		93.3	14.4	15.0	**Gabon**
105		72.6	16.1	10.3	**Gambia**
118		64.2	5.7	21.7	**Georgia**
140		-1.0	1.6	22.3	**Germany**
142		34.0	14.2	10.9	**Ghana**
135		32.8	2.2	24.9	**Greece**
		100.0		21.3	**Grenada**
119		50.3	42.8	21.2	**Guatemala**
		25.9	29.4	7.7	**Guinea**
102	65		28.0	9.5	**Guinea-Bissau**
126		-54.4	9.0	20.2	**Guyana**
89		73.6	20.4	22.7	**Haiti**
117		58.8	19.9	21.4	**Honduras**
135		-84.5		26.4	**Hungary**
144		85.9		21.9	**Iceland**
112	55	-5.2	30.9	3.9	**India**
126	64	13.1	31.8	6.9	**Indonesia**
130	54	37.5	6.3	25.8	**Iran (Islamic Republic of)**
118	61	57.0	11.6	30.4	**Iraq**
152	31	50.5		25.3	**Ireland**
154	33	98.1		26.1	**Israel**
139	34	38.1		19.9	**Italy**
115	38	99.6	8.5	24.7	**Jamaica**
109	41	69.5	5.5	4.3	**Japan**

SELECTED INDICATORS – FOOD SECURITY AND NUTRITION (CONTINUED)

	PREVALENCE OF UNDERNOURISHMENT	NUMBER OF UNDERNOURISHED	PREVALENCE OF SEVERE FOOD INSECURITY	PREVALENCE OF MODERATE OR SEVERE FOOD INSECURITY	AVERAGE DIETARY ENERGY SUPPLY
	PERCENT, 2019–2021	MILLION, 2019–2021	PERCENT, 2019–2021	PERCENT, 2019–2021	KCAL/CAP/DAY, 2019–2021
Jordan	16.9	1.7	17.0	43.0	2 530
Kazakhstan	<2.5		<0.5	2.7	3 346
Kenya	26.9	14.4	26.1	69.5	2 201
Kiribati	4.2	<0.1	8.0	41.0	3 150
Kuwait	2.7	0.1	4.9	12.2	3 371
Kyrgyzstan	5.3	0.3	1.0	6.6	2 728
Lao People's Democratic Republic	5.1	0.4	8.3	31.8	2 772
Latvia	<2.5		0.7	9.4	3 288
Lebanon	10.9	0.7	10.2	29.1	2 871
Lesotho	34.7	0.7	30.9	54.4	2 147
Liberia	38.3	1.9	37.3	80.6	2 158
Libya			20.7	39.4	
Lithuania	<2.5		1.9	9.8	3 455
Luxembourg	<2.5		0.7	2.8	3 452
Madagascar	48.5	13.4	10.3	61.1	1 837
Malawi	17.8	3.4	51.0	81.3	2 670
Malaysia	<2.5		6.3	15.4	2 922
Maldives			2.2	13.4	
Mali	9.8	2.0			2 873
Malta	<2.5		1.4	5.2	3 364
Marshall Islands					
Mauritania	10.1	0.5	7.2	45.3	2 837
Mauritius	7.8	<0.1	9.1	28.2	3 006
Mexico	6.1	7.8	3.7	26.1	3 185

AVERAGE DIETARY ENERGY SUPPLY ADEQUACY	SHARE OF CEREALS/ROOTS/ TUBERS IN DIETARY ENERGY SUPPLY	CEREAL IMPORT DEPENDENCY RATIO	PREVALENCE OF STUNTING, CHILDREN UNDER 5	PREVALENCE OF OBESITY, ADULTS 18 YEARS AND OLDER	
PERCENT, 2019–2021	PERCENT, 2017–2019	PERCENT, 2017–2019	PERCENT, 2020	PERCENT, 2016	
108	43	100.0	7.3	35.5	**Jordan**
144	35	-88.1	6.7	21.0	**Kazakhstan**
99	58	43.1	19.4	7.1	**Kenya**
138			14.9	46.0	**Kiribati**
136	43	96.2	6.0	37.9	**Kuwait**
117	53	17.5	11.4	16.6	**Kyrgyzstan**
118	62	-3.3	30.2	5.3	**Lao People's Democratic Republic**
135	35	-192.7		23.6	**Latvia**
119	44	93.5	10.4	32.0	**Lebanon**
93	68	84.7	32.1	16.6	**Lesotho**
97	68	61.4	28.0	9.9	**Liberia**
		93.9	43.5	32.5	**Libya**
141	37	-189.3		26.3	**Lithuania**
134	29	9.8		22.6	**Luxembourg**
85	79	21.2	40.2	5.3	**Madagascar**
122	64	5.3	37.0	5.8	**Malawi**
123	41	71.5	20.9	15.6	**Malaysia**
			14.2	8.6	**Maldives**
131	69	4.8	25.7	8.6	**Mali**
134	33	88.9		28.9	**Malta**
			32.2	52.9	**Marshall Islands**
126	54		24.2	12.7	**Mauritania**
123	45	90.0	8.7	10.8	**Mauritius**
133	42	35.7	12.1	28.9	**Mexico**

SELECTED INDICATORS – FOOD SECURITY AND NUTRITION (CONTINUED)

	PREVALENCE OF UNDERNOURISHMENT	NUMBER OF UNDERNOURISHED	PREVALENCE OF SEVERE FOOD INSECURITY	PREVALENCE OF MODERATE OR SEVERE FOOD INSECURITY	AVERAGE DIETARY ENERGY SUPPLY
	PERCENT, 2019–2021	MILLION, 2019–2021	PERCENT, 2019–2021	PERCENT, 2019–2021	KCAL/CAP/DAY, 2019–2021
Micronesia (Federated States of)					
Mongolia	3.6	0.1	5.0	25.7	2 951
Montenegro	<2.5		3.4	14.0	3 361
Morocco	5.6	2.1	9.7	31.6	3 365
Mozambique			40.4	73.7	
Myanmar	3.1	1.7	3.7	25.5	2 879
Namibia	18.0	0.5	32.6	57.9	2 569
Nauru					
Nepal	5.5	1.6	13.6	37.8	2 905
Netherlands	<2.5		1.4	4.4	3 345
New Zealand	<2.5		3.5	14.5	3 234
Nicaragua	18.6	1.2			2 659
Niger	19.8	4.8			
Nigeria	12.7	26.2	19.1	58.5	2 603
North Macedonia	3.3	<0.1	6.0	20.9	3 078
Norway	<2.5		1.0	4.3	3 465
Oman	9.8	0.5			2 861
Pakistan	16.9	37.2	8.7	32.6	2 467
Palau					
Panama	5.8	0.2			2 971
Papua New Guinea	21.6	1.9			2 268
Paraguay	8.7	0.6	5.6	25.3	2 770
Peru	8.3	2.7	20.5	50.5	2 834

AVERAGE DIETARY ENERGY SUPPLY ADEQUACY	SHARE OF CEREALS/ROOTS/ TUBERS IN DIETARY ENERGY SUPPLY	CEREAL IMPORT DEPENDENCY RATIO	PREVALENCE OF STUNTING, CHILDREN UNDER 5	PREVALENCE OF OBESITY, ADULTS 18 YEARS AND OLDER	
PERCENT, 2019–2021	PERCENT, 2017–2019	PERCENT, 2017–2019	PERCENT, 2020	PERCENT, 2016	
				45.8	**Micronesia (Federated States of)**
128	37	32.3	7.1	20.6	**Mongolia**
136	32	87.7	8.1	23.3	**Montenegro**
142	60	48.9	12.9	26.1	**Morocco**
		44.4	37.8	7.2	**Mozambique**
125	49	-9.7	25.2	5.8	**Myanmar**
114	55	75.2	18.4	17.2	**Namibia**
			15.0	61.0	**Nauru**
128	64	13.4	30.4	4.1	**Nepal**
130	28	85.2	1.6	20.4	**Netherlands**
130	31	45.9		30.8	**New Zealand**
116	51	38.9	14.1	23.7	**Nicaragua**
119	61	6.6	46.7	5.5	**Niger**
119	67	16.5	35.3	8.9	**Nigeria**
123	35	27.4	4.1	22.4	**North Macedonia**
136	32	43.6		23.1	**Norway**
115	40	85.2	12.2	27.0	**Oman**
110	50	-20.2	36.7	8.6	**Pakistan**
				55.3	**Palau**
129	45	66.8	14.7	22.7	**Panama**
102	46	96.0	48.4	21.3	**Papua New Guinea**
117	48	-70.9	4.6	20.3	**Paraguay**
123	52	54.1	10.8	19.7	**Peru**

SELECTED INDICATORS – FOOD SECURITY AND NUTRITION (CONTINUED)

	PREVALENCE OF UNDERNOURISHMENT	NUMBER OF UNDERNOURISHED	PREVALENCE OF SEVERE FOOD INSECURITY	PREVALENCE OF MODERATE OR SEVERE FOOD INSECURITY	AVERAGE DIETARY ENERGY SUPPLY
	PERCENT, 2019–2021	MILLION, 2019–2021	PERCENT, 2019–2021	PERCENT, 2019–2021	KCAL/CAP/DAY, 2019–2021
Philippines	5.2	5.7	4.8	43.8	2 861
Poland	<2.5		0.9	7.4	3 511
Portugal	<2.5		3.2	11.6	3 465
Qatar					
Republic of Korea	<2.5		0.7	5.3	3 387
Republic of Moldova	6.7	0.3	4.9	24.9	
Romania	<2.5		3.7	13.4	3 610
Russian Federation	<2.5		<0.5	5.5	3 365
Rwanda	35.8	4.6			2 230
Saint Kitts and Nevis			6.4	26.9	
Saint Lucia					
Saint Vincent and the Grenadines	7.6	<0.1	10.3	33.3	2 865
Samoa	4.4	<0.1	3.4	23.6	3 082
San Marino					
Sao Tome and Principe	13.5	<0.1	14.1	54.6	2 395
Saudi Arabia	3.7	1.3			3 302
Senegal	7.5	1.2	11.2	49.2	2 695
Serbia	3.3	0.3	3.8	14.1	2 936
Seychelles					
Sierra Leone	27.4	2.2	31.5	86.7	2 327
Singapore			0.7	4.6	
Slovakia	3.8	0.2	1.6	7.7	2 912
Slovenia	<2.5		0.6	7.4	3 149

AVERAGE DIETARY ENERGY SUPPLY ADEQUACY	SHARE OF CEREALS/ROOTS/ TUBERS IN DIETARY ENERGY SUPPLY	CEREAL IMPORT DEPENDENCY RATIO	PREVALENCE OF STUNTING, CHILDREN UNDER 5	PREVALENCE OF OBESITY, ADULTS 18 YEARS AND OLDER	
PERCENT, 2019–2021	PERCENT, 2017–2019	PERCENT, 2017–2019	PERCENT, 2020	PERCENT, 2016	
127	59	27.2	28.7	6.4	**Philippines**
140	38	-12.5	2.3	23.1	**Poland**
139	32	75.8	3.3	20.8	**Portugal**
			4.6	35.1	**Qatar**
138	35	74.3	2.2	4.7	**Republic of Korea**
107	49	-50.6	4.9	18.9	**Republic of Moldova**
145	40	-40.7	9.7	22.5	**Romania**
137	41	-72.6		23.1	**Russian Federation**
101	52	34.5	32.6	5.8	**Rwanda**
				22.9	**Saint Kitts and Nevis**
		100.0	2.8	19.7	**Saint Lucia**
116	34	96.6		23.7	**Saint Vincent and the Grenadines**
131	32	93.1	6.8	47.3	**Samoa**
					San Marino
106	42		11.8	12.4	**Sao Tome and Principe**
136	49	93.8	3.9	35.4	**Saudi Arabia**
120	63	42.7	17.2	8.8	**Senegal**
118	42	-50.9	5.3	21.5	**Serbia**
		95.6	7.4	14.0	**Seychelles**
104	70	39.8	26.8	8.7	**Sierra Leone**
			2.8	6.1	**Singapore**
115	30	-74.2		20.5	**Slovakia**
127	39	28.2		20.2	**Slovenia**

SELECTED INDICATORS – FOOD SECURITY AND NUTRITION (CONTINUED)

	PREVALENCE OF UNDERNOURISHMENT	NUMBER OF UNDERNOURISHED	PREVALENCE OF SEVERE FOOD INSECURITY	PREVALENCE OF MODERATE OR SEVERE FOOD INSECURITY	AVERAGE DIETARY ENERGY SUPPLY
	PERCENT, 2019–2021	MILLION, 2019–2021	PERCENT, 2019–2021	PERCENT, 2019–2021	KCAL/CAP/DAY, 2019–2021
Solomon Islands	18.1	0.1			2 399
Somalia			41.6	77.4	
South Africa	6.9	4.1	8.0	19.0	2 860
South Sudan			62.3	86.4	
Spain	<2.5		2.0	8.6	3 347
Sri Lanka	3.4	0.7	1.1	10.0	2 833
Sudan	12.8	5.6	17.4	50.7	2 586
Suriname	8.2	<0.1	7.2	35.8	2 720
Sweden	<2.5		1.3	5.3	3 164
Switzerland	<2.5		<0.5	2.2	3 391
Syrian Arab Republic					
Tajikistan	8.6	0.8			2 781
Thailand	8.8	6.2	10.5	33.8	2 830
Timor-Leste	26.2	0.3			2 234
Togo	18.8	1.6	18.8	62.5	2 519
Tonga			6.0	23.2	
Trinidad and Tobago	7.5	0.1	10.2	43.3	2 939
Tunisia	3.1	0.4	12.6	28.0	3 497
Türkiye	<2.5				3 755
Turkmenistan	3.5	0.2			2 883
Tuvalu					
Uganda			23.2	72.5	
Ukraine	2.8	1.2	3.2	22.7	3 036
United Arab Emirates	5.6	0.6	0.8	7.5	3 091

AVERAGE DIETARY ENERGY SUPPLY ADEQUACY	SHARE OF CEREALS/ROOTS/ TUBERS IN DIETARY ENERGY SUPPLY	CEREAL IMPORT DEPENDENCY RATIO	PREVALENCE OF STUNTING, CHILDREN UNDER 5	PREVALENCE OF OBESITY, ADULTS 18 YEARS AND OLDER	
PERCENT, 2019–2021	PERCENT, 2017–2019	PERCENT, 2017–2019	PERCENT, 2020	PERCENT, 2016	
111	66		29.3	22.5	Solomon Islands
			27.4	8.3	Somalia
118	52	9.2	23.2	28.3	South Africa
			30.6		South Sudan
134	26	42.0		23.8	Spain
124	58	34.0	16.0	5.2	Sri Lanka
116	53	25.0	33.7		Sudan
115	44	-17.3	8.0	26.4	Suriname
125	29	-14.1		20.6	Sweden
133	24	53.9		19.5	Switzerland
		36.5	29.6	27.8	Syrian Arab Republic
123	55	48.1	15.3	14.2	Tajikistan
116	46	-57.1	12.3	10.0	Thailand
102	64	42.4	48.8	3.8	Timor-Leste
113	70	19.5	23.8	8.4	Togo
			2.6	48.2	Tonga
121	38	100.0	8.7	18.6	Trinidad and Tobago
149	49	63.1	8.6	26.9	Tunisia
157	44	7.7		32.1	Türkiye
125	57	15.7	7.6	18.6	Turkmenistan
			9.7	51.6	Tuvalu
		6.0	27.9	5.3	Uganda
123	45	-241.9	15.9	24.1	Ukraine
116	36	99.9		31.7	United Arab Emirates

SELECTED INDICATORS – FOOD SECURITY AND NUTRITION (CONTINUED)

	PREVALENCE OF UNDERNOURISHMENT	NUMBER OF UNDERNOURISHED	PREVALENCE OF SEVERE FOOD INSECURITY	PREVALENCE OF MODERATE OR SEVERE FOOD INSECURITY	AVERAGE DIETARY ENERGY SUPPLY
	PERCENT, 2019–2021	MILLION, 2019–2021	PERCENT, 2019–2021	PERCENT, 2019–2021	KCAL/CAP/DAY, 2019–2021
United Kingdom of Great Britain and Northern Ireland	<2.5		1.1	3.5	3 345
United Republic of Tanzania	22.6	13.5	25.8	57.6	2 412
United States of America	<2.5		0.7	8.2	3 864
Uruguay	<2.5		7.3	23.0	3 182
Uzbekistan	<2.5		5.6	23.5	3 270
Vanuatu	11.9	<0.1	2.4	23.3	2 654
Venezuela (Bolivarian Republic of)	22.9	6.5			2 295
Viet Nam	5.7	5.6	0.6	7.6	2 991
Yemen	41.4	12.3			2 099
Zambia	30.9	5.7	32.6	69.5	
Zimbabwe			31.3	73.0	

Source: FAO. 2022. FAOSTAT: Suite of Food Security Indicators. In: *FAO*. Rome. Cited October 2022. https://www.fao.org/faostat/en/#data/FS

AVERAGE DIETARY ENERGY SUPPLY ADEQUACY	SHARE OF CEREALS/ROOTS/ TUBERS IN DIETARY ENERGY SUPPLY	CEREAL IMPORT DEPENDENCY RATIO	PREVALENCE OF STUNTING, CHILDREN UNDER 5	PREVALENCE OF OBESITY, ADULTS 18 YEARS AND OLDER	
PERCENT, 2019–2021	PERCENT, 2017–2019	PERCENT, 2017–2019	PERCENT, 2020	PERCENT, 2016	
134	33	12.2		27.8	**United Kingdom of Great Britain and Northern Ireland**
113	56	4.2	32.0	8.4	**United Republic of Tanzania**
152	23	-23.0	3.2	36.2	**United States of America**
132	38	-95.8	6.5	27.9	**Uruguay**
138	50	26.4	9.9	16.6	**Uzbekistan**
122	47	93.9	28.7	25.2	**Vanuatu**
97	40	60.9	10.6	25.6	**Venezuela (Bolivarian Republic of)**
130	57	8.6	22.3	2.1	**Viet Nam**
96	66	96.3	37.2	17.1	**Yemen**
106	70	-2.8	32.3	8.1	**Zambia**
		24.6	23.0	15.5	**Zimbabwe**

SELECTED INDICATORS – SUSTAINABILITY AND ENVIRONMENT

	SHARE OF FOREST LAND AREA IN TOTAL LAND AREA	SHARE OF AGRICULTURAL LAND AREA IN TOTAL LAND AREA	LAND AREA EQUIPPED FOR IRRIGATION	CROPLAND AREA PER CAPITA	AGRICULTURE AREA UNDER ORGANIC AGRICULTURE
	PERCENT, 2020	PERCENT, 2020	THOUSAND HA, 2020	HA/CAP, 2020	THOUSAND HA, 2020
WORLD	31.2	36.4	348 503	0.20	74 959
AFRICA	21.3	37.6	16 899	0.21	2 087
AMERICAS	41.3	29.0	56 697	0.36	13 857
ASIA	20.0	53.7	245 390	0.13	6 534
EUROPE	46.0	20.9	26 205	0.39	16 573
OCEANIA	21.8	43.3	3 312	0.78	35 908
Afghanistan	1.9	58.8	3 208	0.21	0
Albania	28.8	42.5	360	0.24	1
Algeria	0.8	17.4	1 365	0.19	1
Andorra	34.0	39.8	0	0.01	0
Angola	53.4	45.7	86	0.16	
Antigua and Barbuda	18.5	20.5	0	0.05	
Argentina	10.4	39.6	2 357	0.75	4 400
Armenia	11.5	58.9	209	0.17	1
Australia	17.4	46.3	2 546	1.22	35 688
Austria	47.3	32.1	100	0.15	680
Azerbaijan	13.7	57.8	1 480	0.23	38
Bahamas	50.9	1.4	1	0.03	0
Bahrain	0.9	11.0	4	0.00	
Bangladesh	14.5	76.1	8 127	0.06	1
Barbados	14.7	23.3	5	0.03	
Belarus	43.2	40.8	30	0.61	7
Belgium	22.8	45.1	24	0.08	99
Belize	56.0	7.5	4	0.31	0
Benin	27.8	35.0	24	0.28	39

WATER STRESS	TOTAL RENEWABLE WATER RESOURCES PER CAPITA	EMISSIONS ON AGRICULTURAL LAND	FARM-GATE EMISSIONS	SHARE OF EMISSIONS DUE TO AGRICULTURE AND RELATED LAND USE IN TOTAL EMISSIONS	
PERCENT, 2019	M^3/CAP, 2019	MILLION TONNES CO_2EQ, 2020	MILLION TONNES CO_2EQ, 2020	PERCENT, 2020	
18.6		10 499	7 350	20.2	**WORLD**
		2 427	1 185	50.8	**AFRICA**
		3 112	1 780	29.1	**AMERICAS**
		3 791	3 315	13.4	**ASIA**
		952	869	15.4	**EUROPE**
		217	201	29.2	**OCEANIA**
54.8	1 717	16	16	49.3	Afghanistan
6.8	10 483	3	3	35.8	Albania
137.9	271	14	13	5.1	Algeria
	4 091	0	0		Andorra
1.9	4 663	83	35	54.5	Angola
8.5	535	0	0	4.3	Antigua and Barbuda
10.5	19 567	212	156	54.5	Argentina
57.8	2 627	2	2	20.8	Armenia
3.9	19 521	119	118	21.5	Australia
9.6	8 677	9	9	10.2	Austria
55.6	3 451	10	9	16.3	Azerbaijan
	1 797	0	0	5.5	Bahamas
133.7	71	0	0	0.5	Bahrain
5.7	7 526	121	120	46.5	Bangladesh
87.5	279	0	0	5.0	Barbados
4.5	6 125	48	48	53.1	Belarus
54.1	1 586	12	12	9.7	Belgium
1.3	55 678	6	1	88.2	Belize
1.0	2 236	17	6	59.8	Benin

SELECTED INDICATORS – SUSTAINABILITY AND ENVIRONMENT (CONTINUED)

	SHARE OF FOREST LAND AREA IN TOTAL LAND AREA	SHARE OF AGRICULTURAL LAND AREA IN TOTAL LAND AREA	LAND AREA EQUIPPED FOR IRRIGATION	CROPLAND AREA PER CAPITA	AGRICULTURE AREA UNDER ORGANIC AGRICULTURE
	PERCENT, 2020	PERCENT, 2020	THOUSAND HA, 2020	HA/CAP, 2020	THOUSAND HA, 2020
Bhutan	71.5	13.5	33	0.13	4
Bolivia (Plurinational State of)	46.9	34.9	520	0.41	179
Bosnia and Herzegovina	42.7	43.3	3	0.34	2
Botswana	26.9	45.6	4	0.11	
Brazil	59.4	28.3	8 200	0.30	1 319
Brunei Darussalam	72.1	2.5	1	0.02	
Bulgaria	35.9	46.5	136	0.52	116
Burkina Faso	22.7	44.4	55	0.29	66
Burundi	10.9	79.2	23	0.13	0
Cabo Verde	11.3	19.6	4	0.10	0
Cambodia	45.7	32.8	270	0.26	36
Cameroon	43.0	20.6	29	0.29	0
Canada	38.7	6.4	1 218	1.02	1 418
Central African Republic	35.8	8.2	1	0.39	
Chad	3.4	39.9	30	0.32	
Chile	24.5	21.1	1 110	0.09	157
China	23.3	56.1	74 921	0.09	2 446
Colombia	53.3	43.5	1 087	0.17	51
Comoros	17.7	70.4	0	0.13	1
Congo	64.3	31.1	2	0.11	
Costa Rica	59.4	34.5	160	0.11	11
Côte d'Ivoire	8.9	66.7	73	0.30	79
Croatia	34.7	26.9	48	0.24	109

WATER STRESS	TOTAL RENEWABLE WATER RESOURCES PER CAPITA	EMISSIONS ON AGRICULTURAL LAND	FARM-GATE EMISSIONS	SHARE OF EMISSIONS DUE TO AGRICULTURE AND RELATED LAND USE IN TOTAL EMISSIONS	
PERCENT, 2019	M³/CAP, 2019	MILLION TONNES CO_2EQ, 2020	MILLION TONNES CO_2EQ, 2020	PERCENT, 2020	
1.4	102 216	1	1	28.4	**Bhutan**
1.2	49 856	108	32	55.0	**Bolivia (Plurinational State of)**
2.0	11 360	3	3	9.9	**Bosnia and Herzegovina**
2.1	5 313	48	7	85.4	**Botswana**
1.6	40 971	1 241	567	76.1	**Brazil**
3.5	19 618	1	0	3.4	**Brunei Darussalam**
40.1	3 043	10	7	58.8	**Bulgaria**
7.8	664	32	25	51.9	**Burkina Faso**
10.2	1 087	6	6	65.8	**Burundi**
8.4	546	0	0	14.4	**Cabo Verde**
1.0	28 878	56	23	70.1	**Cambodia**
1.6	10 942	51	16	65.5	**Cameroon**
3.7	77 571	198	102	26.2	**Canada**
0.3	29 714	52	18	94.8	**Central African Republic**
4.3	2 866	106	82	93.7	**Chad**
21.6	48 705	13	13	23.8	**Chile**
43.2	1 938	786	786	5.9	**China**
4.2	46 882	160	77	54.4	**Colombia**
0.8	1 410	0	0	40.0	**Comoros**
0.0	154 632	14	7	42.1	**Congo**
4.1	22 387	4	4	55.8	**Costa Rica**
5.1	3 272	35	11	57.8	**Côte d'Ivoire**
1.5	25 543	4	3	18.8	**Croatia**

SELECTED INDICATORS – SUSTAINABILITY AND ENVIRONMENT (CONTINUED)

	SHARE OF FOREST LAND AREA IN TOTAL LAND AREA	SHARE OF AGRICULTURAL LAND AREA IN TOTAL LAND AREA	LAND AREA EQUIPPED FOR IRRIGATION	CROPLAND AREA PER CAPITA	AGRICULTURE AREA UNDER ORGANIC AGRICULTURE
	PERCENT, 2020	PERCENT, 2020	THOUSAND HA, 2020	HA/CAP, 2020	THOUSAND HA, 2020
Cuba	31.2	61.7	639	0.31	2
Cyprus	18.7	14.5	38	0.11	6
Czechia	34.7	45.7	36	0.24	543
Democratic People's Republic of Korea	50.1	21.5	1 460	0.10	
Democratic Republic of the Congo	55.7	14.8	11	0.17	118
Denmark	15.7	65.5	299	0.41	306
Djibouti	0.3	73.4	1	0.00	
Dominica	63.8	33.3	0	0.32	
Dominican Republic	44.4	50.3	307	0.11	117
Ecuador	50.3	21.8	1 710	0.14	42
Egypt	0.1	4.0	3 928	0.04	116
El Salvador	28.2	57.7	45	0.14	3
Equatorial Guinea	87.3	6.7	1	0.13	
Eritrea	8.7	62.7	21	0.20	
Estonia	57.0	23.0	2	0.53	221
Eswatini	28.9	71.1	50	0.16	1
Ethiopia	15.1	34.1	858	0.16	235
Fiji	62.4	17.1	4	0.15	19
Finland	73.7	7.5	55	0.41	316
France	31.5	52.2	2 691	0.29	2 517
Gabon	91.3	8.6	4	0.22	
Gambia	24.0	59.8	5	0.18	

WATER STRESS	TOTAL RENEWABLE WATER RESOURCES PER CAPITA	EMISSIONS ON AGRICULTURAL LAND	FARM-GATE EMISSIONS	SHARE OF EMISSIONS DUE TO AGRICULTURE AND RELATED LAND USE IN TOTAL EMISSIONS	
PERCENT, 2019	M³/CAP, 2019	MILLION TONNES CO_2EQ, 2020	MILLION TONNES CO_2EQ, 2020	PERCENT, 2020	
23.9	3 363	13	13	33.3	**Cuba**
27.6	651	1	1	7.3	**Cyprus**
29.7	1 230	9	8	8.1	**Czechia**
27.7	3 006	9	6	14.9	**Democratic People's Republic of Korea**
0.2	14 783	660	28	93.7	**Democratic Republic of the Congo**
24.9	1 040	16	14	38.7	**Denmark**
6.3	308	1	1	42.2	**Djibouti**
10.0	2 785	0	0	22.2	**Dominica**
39.6	2 188	11	10	28.8	**Dominican Republic**
6.8	25 464	40	14	44.2	**Ecuador**
141.2	573	27	27	8.1	**Egypt**
2.4	4 071	3	2	28.0	**El Salvador**
0.2	19 174	4	0	18.2	**Equatorial Guinea**
11.2	2 092	6	5	71.5	**Eritrea**
10.9	9 660	5	5	22.3	**Estonia**
77.6	3 928	1	1	66.3	**Eswatini**
32.3	1 089	174	142	77.6	**Ethiopia**
0.3	32 080	1	1	66.4	**Fiji**
7.1	19 884	15	15	20.5	**Finland**
23.5	3 240	90	90	26.1	**France**
0.5	76 407	7	1	30.9	**Gabon**
2.2	3 408	2	1	59.8	**Gambia**

SELECTED INDICATORS – SUSTAINABILITY AND ENVIRONMENT (CONTINUED)

	SHARE OF FOREST LAND AREA IN TOTAL LAND AREA	SHARE OF AGRICULTURAL LAND AREA IN TOTAL LAND AREA	LAND AREA EQUIPPED FOR IRRIGATION	CROPLAND AREA PER CAPITA	AGRICULTURE AREA UNDER ORGANIC AGRICULTURE
	PERCENT, 2020	PERCENT, 2020	THOUSAND HA, 2020	HA/CAP, 2020	THOUSAND HA, 2020
Georgia	40.6	34.2	112	0.11	2
Germany	32.7	47.5	676	0.14	1 593
Ghana	35.1	55.4	223	0.17	75
Greece	30.3	45.5	1 530	0.31	535
Grenada	52.1	23.5	2	0.06	0
Guatemala	32.9	36.0	338	0.11	87
Guinea	25.2	59.0	95	0.29	
Guinea-Bissau	70.4	29.0	25	0.28	10
Guyana	93.6	6.3	143	0.58	
Haiti	12.6	66.8	97	0.12	3
Honduras	56.8	31.4	90	0.16	66
Hungary	22.5	53.7	190	0.43	301
Iceland	0.5	18.6	0	0.35	5
India	24.3	60.2	72 504	0.12	2 658
Indonesia	49.1	33.2	6 722	0.19	76
Iran (Islamic Republic of)	6.6	29.0	9 600	0.21	12
Iraq	1.9	21.3	3 525	0.13	0
Ireland	11.4	65.5		0.09	75
Israel	6.5	29.9	306	0.06	5
Italy	32.4	44.0	4 124	0.15	2 095
Jamaica	55.1	41.0	31	0.07	0
Japan	68.4	12.0	2 405	0.03	12
Jordan	1.1	11.6	107	0.03	1
Kazakhstan	1.3	79.3	2 234	1.58	115

WATER STRESS	TOTAL RENEWABLE WATER RESOURCES PER CAPITA	EMISSIONS ON AGRICULTURAL LAND	FARM-GATE EMISSIONS	SHARE OF EMISSIONS DUE TO AGRICULTURE AND RELATED LAND USE IN TOTAL EMISSIONS	
PERCENT, 2019	M^3/CAP, 2019	MILLION TONNES CO_2EQ, 2020	MILLION TONNES CO_2EQ, 2020	PERCENT, 2020	
5.1	15 845	2	2	14.6	**Georgia**
33.5	1 844	77	77	11.0	**Germany**
6.3	1 848	13	13	63.4	**Ghana**
20.5	6 531	9	9	12.6	**Greece**
7.1	1 786	0	0	11.6	**Grenada**
5.7	7 275	16	12	38.2	**Guatemala**
1.4	17 696	35	24	80.9	**Guinea**
1.5	16 346	4	2	74.4	**Guinea-Bissau**
3.3	346 208	17	7	88.0	**Guyana**
13.4	1 245	5	5	32.8	**Haiti**
4.6	9 456	14	8	49.1	**Honduras**
7.7	10 739	16	16	22.2	**Hungary**
0.4	501 429	1	1	29.9	**Iceland**
66.5	1 398	790	789	21.5	**India**
29.7	7 459	772	513	46.3	**Indonesia**
81.3	1 653	51	51	4.9	**Iran (Islamic Republic of)**
79.5	2 286	12	12	3.1	**Iraq**
20.0	10 650	26	26	44.2	**Ireland**
100.4	209	3	3	3.3	**Israel**
30.0	3 159	44	44	12.2	**Italy**
12.5	3 671	4	4	38.1	**Jamaica**
36.4	3 390	46	44	3.8	**Japan**
104.3	93	2	2	6.1	**Jordan**
32.7	5 844	32	32	9.2	**Kazakhstan**

SELECTED INDICATORS – SUSTAINABILITY AND ENVIRONMENT (CONTINUED)

	SHARE OF FOREST LAND AREA IN TOTAL LAND AREA	SHARE OF AGRICULTURAL LAND AREA IN TOTAL LAND AREA	LAND AREA EQUIPPED FOR IRRIGATION	CROPLAND AREA PER CAPITA	AGRICULTURE AREA UNDER ORGANIC AGRICULTURE
	PERCENT, 2020	PERCENT, 2020	THOUSAND HA, 2020	HA/CAP, 2020	THOUSAND HA, 2020
Kenya	6.3	48.6	151	0.12	124
Kiribati	1.5	42.0		0.28	
Kuwait	0.4	8.4	18	0.00	0
Kyrgyzstan	6.9	54.1	1 023	0.21	30
Lao People's Democratic Republic	71.9	8.8	441	0.19	3
Latvia	54.8	31.6	1	0.71	291
Lebanon	14.0	65.5	90	0.04	2
Lesotho	1.1	85.6	3	0.28	
Liberia	79.1	20.3	3	0.14	
Libya	0.1	8.7	400	0.30	
Lithuania	35.2	47.0	4	0.84	236
Luxembourg	34.5	51.3		0.10	6
Madagascar	21.4	70.3	1 086	0.13	104
Malawi	23.8	59.9	91	0.20	0
Malaysia	58.2	26.1	442	0.26	1
Maldives	2.7	21.3		0.01	
Mali	10.9	33.8	380	0.32	15
Malta	1.4	32.4	4	0.02	0
Marshall Islands	52.2	47.8		0.14	
Mauritania	0.3	38.5	45	0.09	
Mauritius	19.1	42.4	19	0.06	0
Mexico	33.8	50.0	7 304	0.18	216
Micronesia (Federated States of)	92.0	31.4		0.17	

WATER STRESS	TOTAL RENEWABLE WATER RESOURCES PER CAPITA	EMISSIONS ON AGRICULTURAL LAND	FARM-GATE EMISSIONS	SHARE OF EMISSIONS DUE TO AGRICULTURE AND RELATED LAND USE IN TOTAL EMISSIONS	
PERCENT, 2019	M³/CAP, 2019	MILLION TONNES CO_2EQ, 2020	MILLION TONNES CO_2EQ, 2020	PERCENT, 2020	
33.2	584	55	55	62.7	**Kenya**
	0	0	0	18.1	**Kiribati**
3850.5	5	2	2	1.1	**Kuwait**
50.0	3 681	6	6	35.5	**Kyrgyzstan**
4.8	46 517	27	11	55.0	**Lao People's Democratic Republic**
1.1	18 324	6	6	63.0	**Latvia**
58.8	657	1	1	4.6	**Lebanon**
2.6	1 422	1	1	31.4	**Lesotho**
0.3	46 989	14	1	79.2	**Liberia**
817.1	103	4	4	6.3	**Libya**
1.8	8 878	11	11	41.5	**Lithuania**
4.1	5 684	1	1	8.2	**Luxembourg**
11.3	12 496	36	31	73.2	**Madagascar**
17.5	928	17	10	72.8	**Malawi**
3.4	18 153	94	61	22.9	**Malaysia**
15.7	57	0	0	6.3	**Maldives**
8.0	6 104	39	39	82.0	**Mali**
81.2	115	0	0	4.0	**Malta**
	0	0	0	14.0	**Marshall Islands**
13.2	2 519	11	11	77.6	**Mauritania**
21.6	2 167	0	0	3.4	**Mauritius**
44.7	3 621	131	115	17.7	**Mexico**
	0	0	0	50.8	**Micronesia (Federated States of)**

SELECTED INDICATORS – SUSTAINABILITY AND ENVIRONMENT (CONTINUED)

	SHARE OF FOREST LAND AREA IN TOTAL LAND AREA	SHARE OF AGRICULTURAL LAND AREA IN TOTAL LAND AREA	LAND AREA EQUIPPED FOR IRRIGATION	CROPLAND AREA PER CAPITA	AGRICULTURE AREA UNDER ORGANIC AGRICULTURE
	PERCENT, 2020	PERCENT, 2020	THOUSAND HA, 2020	HA/CAP, 2020	THOUSAND HA, 2020
Mongolia	9.1	72.4	87	0.41	0
Montenegro	61.5	19.2	2	0.02	5
Morocco	12.9	68.1	1 946	0.25	11
Mozambique	46.7	52.7	118	0.19	14
Myanmar	43.7	19.9	2 295	0.23	10
Namibia	8.1	47.1	8	0.32	
Nauru	0.0	20.0		0.04	
Nepal	41.6	28.8	1 369	0.08	9
Netherlands	11.0	53.9	522	0.06	72
New Zealand	37.6	38.6	751	0.12	79
Nicaragua	28.3	42.1	199	0.27	39
Niger	0.9	36.8	267	0.74	
Nigeria	23.8	76.3	331	0.20	55
North Macedonia	39.7	50.0	128	0.22	4
Norway	33.4	2.7	85	0.15	45
Oman	0.0	4.7	108	0.02	0
Pakistan	4.8	47.6	19 990	0.14	70
Palau	90.0	9.4		0.13	
Panama	56.8	29.3	41	0.15	6
Papua New Guinea	79.2	2.6		0.11	72
Paraguay	40.5	42.3	140	0.68	73
Peru	56.5	19.1	2 600	0.17	560
Philippines	24.1	42.5	2 006	0.10	192
Poland	31.0	47.2	271	0.30	509

WATER STRESS	TOTAL RENEWABLE WATER RESOURCES PER CAPITA	EMISSIONS ON AGRICULTURAL LAND	FARM-GATE EMISSIONS	SHARE OF EMISSIONS DUE TO AGRICULTURE AND RELATED LAND USE IN TOTAL EMISSIONS	
PERCENT, 2019	M³/CAP, 2019	MILLION TONNES CO_2EQ, 2020	MILLION TONNES CO_2EQ, 2020	PERCENT, 2020	
3.4	10 790	33	33	36.3	**Mongolia**
		0	0	9.9	**Montenegro**
50.8	795	18	18	18.2	**Morocco**
1.8	7 149	77	18	66.4	**Mozambique**
5.8	21 608	216	116	80.4	**Myanmar**
0.9	15 999	21	10	83.0	**Namibia**
	930	0	0	12.3	**Nauru**
8.3	7 347	32	32	59.0	**Nepal**
17.0	5 323	28	28	14.5	**Netherlands**
8.0	68 366	48	46	63.7	**New Zealand**
2.7	25 135	33	12	83.2	**Nicaragua**
11.0	1 461	36	35	82.4	**Niger**
9.7	1 424	138	91	28.7	**Nigeria**
25.3	3 072	1	1	13.6	**North Macedonia**
2.0	73 064	7	7	11.3	**Norway**
116.7	281	3	3	2.7	**Oman**
108.6	1 140	220	213	39.2	**Pakistan**
	0	0	0	3.1	**Palau**
0.9	32 805	8	5	41.2	**Panama**
0.1	91 271	47	35	74.9	**Papua New Guinea**
1.8	55 045	84	35	92.1	**Paraguay**
7.2	57 821	121	28	64.8	**Peru**
26.3	4 430	67	67	26.8	**Philippines**
31.1	1 597	62	62	16.1	**Poland**

SELECTED INDICATORS – SUSTAINABILITY AND ENVIRONMENT (CONTINUED)

	SHARE OF FOREST LAND AREA IN TOTAL LAND AREA	SHARE OF AGRICULTURAL LAND AREA IN TOTAL LAND AREA	LAND AREA EQUIPPED FOR IRRIGATION	CROPLAND AREA PER CAPITA	AGRICULTURE AREA UNDER ORGANIC AGRICULTURE
	PERCENT, 2020	PERCENT, 2020	THOUSAND HA, 2020	HA/CAP, 2020	THOUSAND HA, 2020
Portugal	36.2	42.3	567	0.18	320
Qatar	0.0	6.4	24	0.01	
Republic of Korea	64.4	16.6	707	0.03	39
Republic of Moldova	11.8	68.9	218	0.48	28
Romania	30.1	59.1	3 166	0.49	469
Russian Federation	49.8	13.2	4 300	0.85	615
Rwanda	11.2	73.4	9	0.11	5
Saint Kitts and Nevis	42.3	23.1	0	0.10	
Saint Lucia	34.1	16.3	3	0.05	
Saint Vincent and the Grenadines	73.2	18.0	1	0.05	
Samoa	58.2	17.8		0.22	41
San Marino	16.7	38.3		0.07	
Sao Tome and Principe	54.1	45.8	10	0.20	9
Saudi Arabia	0.5	80.8	3 279	0.10	27
Senegal	41.9	46.1	120	0.20	4
Serbia	31.1	40.1	63	0.32	21
Seychelles	73.3	3.4	0	0.02	
Sierra Leone	35.1	54.7	30	0.22	220
Singapore	21.7	0.9		0.00	0
Slovakia	40.1	39.2	57	0.25	223
Slovenia	61.5	30.3	7	0.11	52
Solomon Islands	90.1	4.2		0.16	3
Somalia	9.5	70.3	200	0.07	

WATER STRESS	TOTAL RENEWABLE WATER RESOURCES PER CAPITA	EMISSIONS ON AGRICULTURAL LAND	FARM-GATE EMISSIONS	SHARE OF EMISSIONS DUE TO AGRICULTURE AND RELATED LAND USE IN TOTAL EMISSIONS	
PERCENT, 2019	M³/CAP, 2019	MILLION TONNES CO_2EQ, 2020	MILLION TONNES CO_2EQ, 2020	PERCENT, 2020	
12.3	7 569	9	9	10.1	Portugal
431.0	20	1	1	0.7	Qatar
85.2	1 361	22	19	3.5	Republic of Korea
12.5	3 035	2	2	20.7	Republic of Moldova
6.0	10 948	23	18	27.9	Romania
4.1	31 023	186	151	9.6	Russian Federation
20.2	1 053	6	6	62.2	Rwanda
50.8	454	0	0	3.3	Saint Kitts and Nevis
14.3	1 641	0	0	12.7	Saint Lucia
7.9	904	0	0	6.8	Saint Vincent and the Grenadines
	0	0	0	41.4	Samoa
		0	0		San Marino
1.9	10 137	0	0	54.9	Sao Tome and Principe
974.2	70	19	19	2.5	Saudi Arabia
16.3	2 391	18	14	53.1	Senegal
6.3	18 490	14	7	17.1	Serbia
	0	0	0	6.1	Seychelles
0.5	20 478	7	3	63.8	Sierra Leone
82.2	103	1	1	1.1	Singapore
2.4	9 181	3	3	6.8	Slovakia
6.4	15 332	3	2	19.0	Slovenia
	66 734	0	0	0.7	Solomon Islands
24.5	952	41	24	83.7	Somalia

SELECTED INDICATORS – SUSTAINABILITY AND ENVIRONMENT (CONTINUED)

	SHARE OF FOREST LAND AREA IN TOTAL LAND AREA	SHARE OF AGRICULTURAL LAND AREA IN TOTAL LAND AREA	LAND AREA EQUIPPED FOR IRRIGATION	CROPLAND AREA PER CAPITA	AGRICULTURE AREA UNDER ORGANIC AGRICULTURE
	PERCENT, 2020	PERCENT, 2020	THOUSAND HA, 2020	HA/CAP, 2020	THOUSAND HA, 2020
South Africa	14.1	79.4	1 670	0.21	41
South Sudan	11.3	44.7	19	0.22	
Spain	37.2	52.3	3 923	0.36	2 438
Sri Lanka	34.2	45.5	637	0.11	73
Sudan	9.8	37.2	1 855	0.48	
Suriname	97.4	0.5	65	0.11	0
Sweden	68.7	7.4	157	0.25	611
Switzerland	32.1	38.1	52	0.05	176
Syrian Arab Republic	2.8	75.8	1 310	0.33	
Tajikistan	3.1	35.4	817	0.11	12
Thailand	38.9	45.0	6 415	0.32	161
Timor-Leste	61.9	23.0	35	0.15	32
Togo	22.2	70.2	8	0.34	128
Tonga	12.4	48.6		0.29	1
Trinidad and Tobago	44.5	10.5	7	0.03	
Tunisia	4.5	62.6	495	0.42	297
Türkiye	28.9	49.1	5 215	0.27	383
Turkmenistan	8.8	72.0	1 995	0.33	
Tuvalu	33.3	60.0		0.15	
Uganda	11.7	71.9	11	0.20	116
Ukraine	16.7	71.3	2 166	0.77	462
United Arab Emirates	4.5	5.5	90	0.01	5

WATER STRESS	TOTAL RENEWABLE WATER RESOURCES PER CAPITA	EMISSIONS ON AGRICULTURAL LAND	FARM-GATE EMISSIONS	SHARE OF EMISSIONS DUE TO AGRICULTURE AND RELATED LAND USE IN TOTAL EMISSIONS	
PERCENT, 2019	M^3/CAP, 2019	MILLION TONNES CO_2EQ, 2020	MILLION TONNES CO_2EQ, 2020	PERCENT, 2020	
63.6	877	44	37	7.4	**South Africa**
4.2	4 475	61	58	87.7	**South Sudan**
40.2	2 386	52	51	19.1	**Spain**
90.8	2 476	9	9	22.8	**Sri Lanka**
118.7	883	101	80	73.0	**Sudan**
4.0	170 287	10	2	78.1	**Suriname**
3.4	17 337	38	14	85.7	**Sweden**
6.5	6 227	7	6	18.1	**Switzerland**
124.4	984	8	7	15.6	**Syrian Arab Republic**
69.9	2 351	7	7	44.1	**Tajikistan**
23.0	6 300	99	83	22.0	**Thailand**
28.3	6 353	1	1	22.0	**Timor-Leste**
3.4	1 819	5	4	38.7	**Togo**
	0	0	0	26.9	**Tonga**
20.3	2 753	0	0	0.8	**Trinidad and Tobago**
96.0	395	6	6	18.3	**Tunisia**
45.7	2 536	68	68	12.5	**Türkiye**
143.6	4 168	13	13	9.1	**Turkmenistan**
	0	0	0	32.3	**Tuvalu**
5.8	1 358	49	39	72.9	**Uganda**
13.7	3 984	47	46	18.1	**Ukraine**
1672.0	15	4	4	1.7	**United Arab Emirates**

SELECTED INDICATORS – SUSTAINABILITY AND ENVIRONMENT (CONTINUED)

	SHARE OF FOREST LAND AREA IN TOTAL LAND AREA	SHARE OF AGRICULTURAL LAND AREA IN TOTAL LAND AREA	LAND AREA EQUIPPED FOR IRRIGATION	CROPLAND AREA PER CAPITA	AGRICULTURE AREA UNDER ORGANIC AGRICULTURE
	PERCENT, 2020	PERCENT, 2020	THOUSAND HA, 2020	HA/CAP, 2020	THOUSAND HA, 2020
United Kingdom of Great Britain and Northern Ireland	13.2	71.3	208	0.09	470
United Republic of Tanzania	51.6	44.6	364	0.26	198
United States of America	33.9	44.4	26 916	0.48	2 327
Uruguay	11.6	80.4	263	0.59	2 742
Uzbekistan	8.4	58.3	4 329	0.13	4
Vanuatu	36.3	15.3		0.47	2
Venezuela (Bolivarian Republic of)	52.4	24.4	1 055	0.12	1
Viet Nam	46.7	39.4	4 585	0.12	64
Yemen	1.0	44.4	680	0.05	
Zambia	60.3	32.1	156	0.21	1
Zimbabwe	45.1	41.9	187	0.28	1

Source: FAO. 2022. FAOSTAT: Land use indicators. In: *FAO*. Rome. Cited October 2022. http://www.fao.org/faostat/en/#data/EL; FAO. 2022. FAOSTAT: Land Use. In: *FAO*. Rome. Cited October 2022. http://www.fao.org/faostat/en/#data/RL; FAO. 2022. AQUASTAT. In: *FAO*. Rome. Cited October 2022. https://www.fao.org/aquastat/statistics/query/index.html?lang=en;
FAO. 2022. FAOSTAT: Emissions Totals. In: *FAO*. Rome. Cited October 2022. https://www.fao.org/faostat/en/#data/GT;
FAO. 2022. FAOSTAT: Emissions shares. In: *FAO*. Rome. Cited October 2022. https://www.fao.org/faostat/en/#data/EM

WATER STRESS	TOTAL RENEWABLE WATER RESOURCES PER CAPITA	EMISSIONS ON AGRICULTURAL LAND	FARM-GATE EMISSIONS	SHARE OF EMISSIONS DUE TO AGRICULTURE AND RELATED LAND USE IN TOTAL EMISSIONS	
PERCENT, 2019	M^3/CAP, 2019	MILLION TONNES CO_2EQ, 2020	MILLION TONNES CO_2EQ, 2020	PERCENT, 2020	
14.4	2 177	55	55	14.2	**United Kingdom of Great Britain and Northern Ireland**
13.0	1 660	133	70	78.3	**United Republic of Tanzania**
28.2	9 326	540	481	9.6	**United States of America**
9.8	49 744	28	28	68.2	**Uruguay**
168.9	1 482	37	37	19.0	**Uzbekistan**
	33 346	0	0	54.7	**Vanuatu**
7.5	46 465	85	44	38.5	**Venezuela (Bolivarian Republic of)**
18.1	9 165	81	81	20.5	**Viet Nam**
169.8	72	8	8	23.0	**Yemen**
2.8	5 868	76	38	81.6	**Zambia**
35.4	1 366	21	11	18.6	**Zimbabwe**

DEFINITIONS AND NOTES

ADULT OBESITY, PREVALENCE

The prevalence of obesity in the adult population is the percentage of adults age 18 and over whose body mass index (BMI) is more than 30 kg/m^2. The BMI is a simple index of weight-for-height, or the weight in kilograms divided by the square of the height in metres.

Source: WHO
Owner: World Health Organization, Global Health Observatory Data Repository/World Health Statistics

AGRICULTURAL LAND

Land used for cultivation of crops and animal husbandry. It is the total of areas under "Cropland" and "Permanent meadows and pastures".

Source: FAO, Statistics Division
Owner: FAO

AGRICULTURE, FORESTRY AND FISHING

Agriculture, forestry and fishing (AFF) refers to the broad agricultural sector including crop growing and animal production, forestry and logging, and fishing and aquaculture. These sub-sectors correspond to Section A of the International Standard Industrial Classification (ISIC), revision 4 and are covered in its Divisions 1, 2 and 3. To distinguish the agricultural sector (crop and livestock in Division 1 of ISIC), the broad agricultural sector is abbreviated as AFF.

Source: United Nations Statistics Division
Owner: UN

AGRICULTURE, FORESTRY AND OTHER LAND USE EMISSIONS

Greenhouse gas (GHG) emissions from agriculture, forestry and other land use (AFOLU) consist of non-CO_2 gases, namely methane (CH_4) and nitrous oxide (N_2O) produced by crop and livestock production and management activities, CO_2 emissions by sources and sinks from forestland, net forest conversion and drained organic soils, and non-CO_2 emissions from forest fires and fires in organic soils.

Source: FAO, Statistics Division
Owner: FAO

AGRICULTURE, VALUE ADDED

This is the total value added in AFF. The value added is the net output of a sector after adding up the value of all outputs and subtracting intermediate inputs. It is calculated without making deductions for depreciation of fabricated assets or depletion and degradation of natural resources. ISIC, revision 3 or 4, determines the origin of value added. Agriculture here refers to the broad agricultural sector (AFF).

Source: World Bank
Owner: World Bank

AID DISBURSEMENT FLOWS TO AGRICULTURE, FORESTRY AND FISHING

The release of funds to or the purchase of goods or services for a recipient; by extension, the amount thus spent. Disbursements record the actual international transfer of financial resources, or of goods or services valued at the cost to the donor. In the case of activities carried out in donor countries, such as training, administration or public awareness programmes, disbursement is taken to have occurred when the funds have been transferred to the service provider or the recipient. They may be recorded gross (the total amount disbursed over a given accounting period) or net (the gross amount less any repayments of loan principal or recoveries on grants received during the same period). It can take several years to disburse a commitment.
The OECD Development Assistance Committee (DAC) uses a sector classification specifically developed to track aid flows and to permit measuring the share of each sector (e.g. health, energy, agriculture) or other purpose category "non-sector allocable aid" (e.g. general budget support, humanitarian aid) in total aid. The sector of destination is assigned by answering the question "which specific area of the recipient's economic and social structure is the transfer intended to foster".

Source: OECD
Owner: OECD

ANIMAL OILS AND FATS

Animal oils and fats include animal fats that are obtained by dressing the carcasses of slaughtered animals (slaughter fats), or at a later stage in the butchering process when meat is being prepared for final consumption (butcher fats).

Source: FAO, Statistics Division
Owner: FAO

AQUACULTURE FISH PRODUCTION

Aquaculture fish production is defined as the farming of aquatic organisms. Farming implies some form of intervention in the rearing

process to enhance production, such as regular stocking, feeding, protection from predators, etc. Farming also implies individual or corporate ownership of the stock being cultivated. For statistical purposes, aquatic organisms, which are harvested by an individual or corporate body that has owned them throughout their rearing period, contribute to aquaculture, while aquatic organisms, which are exploitable by the public as a common property resource, with or without appropriate licences, are the harvest of fisheries.
In the case of capture-based aquaculture, only the incremental growth (or weight gain) in captivity, could and should be reported as the production from aquaculture. Data included here cover aquaculture production of fish, molluscs, crustaceans and miscellaneous aquatic animals but exclude the production of marine mammals, crocodiles, corals, pearls, sponges and algae. Fish production is the sum of aquaculture and capture fish production. Data are expressed in live weight equivalent.

Source: FAO, Fisheries and Aquaculture Division
Owner: FAO

ARABLE LAND

Arable land is the total of areas under temporary crops, temporary meadows and pastures, and land with temporary fallow. Arable land does not include land that is potentially cultivable but is not normally cultivated.

Source: FAO, Statistics Division
Owner: FAO

CAPTURE FISH PRODUCTION

Hunting, collecting and gathering activities directed at removing or collecting live wild aquatic organisms are capture fish production. The capture production statistics here indicates the nominal catches of aquatic organisms, killed, caught, trapped or collected for all commercial, industrial, recreational and subsistence purposes or other utilizations in live weight equivalent. Data included here cover capture production of fish, molluscs, crustaceans and miscellaneous aquatic animals but exclude production of marine mammals, crocodiles, corals, pearls, sponges and algae.
Fish production is the sum of aquaculture and capture fish production.

Source: FAO, Fisheries and Aquaculture Division
Owner: FAO

CEREAL IMPORT DEPENDENCY RATIO

The cereal import dependency ratio provides a measure of the dependence of a country or region from cereal imports. The higher the value of the indicator, the higher the dependence. Specifically, the cereal imports dependency ratio tells how much of the available domestic food supply of cereals has been imported and how much comes from the country's own production. It is computed as (cereal imports - cereal exports) / (cereal production + cereal imports - cereal exports) x 100. Given this formula the indicator assumes only values less than or equal to 100. Negative values indicate that the country is a net exporter of cereals. The indicator is calculated in three-year averages, to reduce the impact of possible errors in estimated production and trade, due to the difficulties in properly accounting for stock variations in major foods.

Source: FAO, Statistics Division
Owner: FAO

CEREALS

Wheat, rice paddy, barley, maize, popcorn, rye, oats, millets, sorghum, buckwheat, quinoa, fonio, triticale, canary seed, mixed grain and cereals nes are all considered cereals.

Source: FAO, Statistics Division
Owner: FAO

CEREALS AND PREPARATIONS

Cereals, flours and cereal grains that are either rolled, flaked, pearled, sliced or kibbled are cereals and preparations.

Source: FAO, Statistics Division
Owner: FAO

CONSUMER PRICE INDEX (FOOD)

The food consumer price index (CPI) measures the price change between the current and reference periods of the average basket of food items purchased by households. The food CPI is rescaled to a unique base year of 2010 by FAO for all countries with sufficient time coverage. FAO uses the geometric mean of the monthly indices of the year 2010 as the rescaling factor.

Source: IMF, UNSD, OECD and national statistics' websites
Owner: IMF, UNSD and FAO

CROPLAND

Cropland is the land used for cultivation of crops. The total of areas under "Arable land" and "Permanent crops".

Source: FAO, Statistics Division
Owner: FAO

CROPS

Crop statistics include permanent and temporary crops and cover the following categories: Crops primary, Fibre crops primary, Cereals, Coarse grain, Citrus fruit, Fruit, Oil crops (oil and cake equivalent), Pulses, Roots and tubers, Treenuts and Vegetables.

Source: FAO, Statistics Division
Owner: FAO

DAIRY PRODUCTS

Butter, buttermilk, cheese, cream, ghee, milk, whey and yoghurt are all dairy products.

Source: FAO, Statistics Division
Owner: FAO

DIETARY ENERGY SUPPLY (KCAL/CAP/DAY)

The food available for human consumption, expressed in kilocalories per person per day is the dietary energy supply. At the country level, it is calculated as the food remaining for human use after taking out all non-food utilization, including exports, industrial use, animal feed, seed, wastage and changes in stocks.

Source: FAO, Statistics Division
Owner: FAO

DIETARY ENERGY SUPPLY, AVERAGE

The figures for the dietary energy supply average are based on the latest available data from national food balance sheets and represent the amount of food available for human consumption.

Source: FAO, Statistics Division
Owner: FAO

EGGS, PRIMARY

Egg production by type of poultry should refer to the total production of eggs in the shell by all types of hens in both the traditional sector (individually owned small flocks) and the modern sector (large-scale, intensive commercial poultry farms). Total production includes eggs for hatching but excludes farm waste.

Source: FAO, Statistics Division
Owner: FAO

EMISSIONS ON AGRICULTURAL LAND

Emissions on agricultural land are composed of the sum of emissions within the farm gate and food-related land use change emissions from net forest conversion, fires in organic soils and fires in humid tropical forests.

Source: FAO, Statistics Division
Owner: FAO

EMISSIONS SHARES

Emissions from the different economic sectors (energy, agriculture, land use, land-use change and forestry [LULUCF], industrial processes and product use, waste and international bunkers) and their related contributions to all emissions by gases (CO_2, CH_4, N_2O and F-gases) are computed in this domain. Emissions from food systems that can be associated with farm gate activities, food-related land use change and pre- and post- production activities are also computed along with their contributions to all emissions.

Source: FAO, Statistics Division
Owner: FAO

EMPLOYMENT IN AGRICULTURE (TOTAL POPULATION, MALE, FEMALE)

Employment comprises all persons of working age who, during a specified brief period, such as one week or one day, were in the following categories: a) paid employment (whether at work or having a job but not at work); or b) self-employment (whether at work or with an enterprise but not at work). The working-age population is the population above the legal working age, but for statistical purposes it comprises all persons above a specified minimum age threshold for which an inquiry on economic activity is made. To promote international comparability, the working-age population is often defined as all persons aged 15 and older, but this may vary from country to country based on national laws and practices (some countries also use an upper age limit). The classification by economic activity refers to the main activity of the establishment in which a person worked during the reference period. The branch of economic activity of a person does not depend on the specific duties or functions of the person's job, but rather on the characteristics of the economic unit in which the person worked. Data presented by a branch of economic activity are based on ISIC.

Source: ILO
Owner: ILO

EXPORT VALUE

Export values are reported as FOB (free on board: the value of the goods plus the value of the services performed to deliver the goods to the border of the exporting country).

Source: FAO, Statistics Division
Owner: FAO

FARM-GATE EMISSIONS

Farm-gate emissions covers all GHG emissions produced from agricultural processes (enteric fermentation, manure management, rice cultivation, synthetic fertilizers, manure applied to soils, manure left on pastures, crop residues, drained organic soils, burning of crop residues, savanna fires, energy use) within the farm gate and at the farm boundary. Non-CO_2 gases, namely methane (CH_4) and nitrous oxide (N_2O), are produced by crop and livestock production and management activities whereas CO_2 emissions are produced from the drainage of organic soils and energy used on farms. The FAOSTAT emissions database is computed following Tier 1 Intergovernmental Panel on Climate Change (IPCC) 2006 Guidelines for National GHG Inventories.

Source: FAO, Statistics Division
Owner: FAO

FERTILIZERS

The use of fertilizers refers to agricultural use of inorganic (mineral or chemical) fertilizers for the three main plant nutrients: nitrogen (N), phosphorus (expressed as P_2O_5) and potassium (expressed as K_2O). It includes both straight fertilizers (those containing only one of the three primary plant nutrients) and compound fertilizers (those containing more than one of the three primary plant nutrients; they may be NP, NK, PK or NPK). Agricultural use refers to the use for crops, livestock, forestry, fisheries and aquaculture, excluding use for animal feed.

Source: FAO, Statistics Division
Owner: FAO

FISH NET TRADE

Fish net trade is exports plus re-exports minus imports.

Source: FAO, Fisheries and Aquaculture Division
Owner: FAO

FOOD

Food is comprised of the commodities in the Standard International Trade Classification (SITC) sections 0 (food and live animals), 1 (beverages and tobacco), and 4 (animal and vegetable oils and fats) and SITC division 22 (oil seeds, oil nuts and oil kernels).

Source: FAO, Statistics Division
Owner: FAO

FOREST LAND

Land spanning more than 0.5 ha with trees higher than 5 metres

and a canopy cover of more than 10 percent, or trees able to reach these thresholds *in situ*. Excludes land that is predominantly under agricultural or urban land use.
Explanatory notes:

- Forest land is determined both by the presence of trees and by the absence of other predominant land uses. The trees should be able to reach a minimum height of 5 metres *in situ*.
- Includes areas with young trees that have not yet reached but that are expected to reach a canopy cover of 10 percent and tree height of 5 metres. It also includes areas that are temporarily unstocked owing to clear-cutting as part of a forest management practice or natural disasters, and that are expected to be regenerated within five years. Local conditions may, in exceptional cases, justify the use of a longer time frame.
- Includes forest roads, firebreaks and other small open areas.
- May include forest land in national parks, nature reserves and other protected areas, such as those of specific environmental, scientific, historical, cultural or spiritual interest.
- Includes windbreaks, shelter belts and corridors of trees with an area of more than 0.5 ha and width of more than 20 metres.
- Includes abandoned shifting cultivation land with a regeneration of trees that have, or are expected to reach, a canopy cover of 10 percent and tree height of 5 metres.
- Includes areas with mangroves in tidal zones, regardless of whether this area is classified as land area or not.
- Includes areas with bamboo and palms provided that land use, height and canopy cover criteria are met.
- Some agroforestry systems such as the taungya system, where crops are grown only during the first years of the forest rotation, should be classified as forest.
- Excludes tree stands in agricultural production systems, such as fruit-tree plantations (permanent crops), oil palm plantations, rubber and Christmas trees (permanent crops) and agroforestry systems when crops are grown under tree cover.

Source: FAO, Statistics Division
Owner: FAO

FOREST PRODUCTS EXPORTS

Products of domestic origin or manufacture shipped out of the country are forest product exports. They include exports from free economic zones and re-exports and exclude "in-transit" shipments. They are reported in cubic metres of solid volume or metric tonnes and values are normally recorded as FOB.

Source: FAO, Forestry Division
Owner: FAO

FOREST PRODUCTS IMPORTS

Products imported for domestic consumption or processing shipped into a country are forest product imports. They include imports into free economic zones or for re-export and exclude "in-transit" shipments. They are reported in cubic metres of solid volume or metric tonnes and values normally include cost, insurance and freight (CIF).

Source: FAO, Forestry Division
Owner: FAO

FOREST PRODUCTS PRODUCTION

Forest product production includes the production of products that may immediately be consumed in the production of another product (e.g. wood pulp, which may immediately be converted into paper as part of a continuous process). This includes production from all sources within the country including public, private and informal sources. It excludes the production of veneer sheets that are used for plywood production within the same country. It is reported in cubic metres of solid volume in the case of roundwood, sawnwood and wood-based panels and metric tonnes in the case of charcoal, pulp and paper products.

Source: FAO, Forestry Division
Owner: FAO

FRUIT AND VEGETABLES

Vegetables, as classified in this group, are mainly annual plants cultivated as field and garden crops in the open and under glass and are used almost exclusively for food. Vegetables grown principally for animal feed or seed should be excluded. Certain plants, normally classified as cereals and pulses, belong to this group when harvested green, such as green maize, green peas, etc. Chilies and green peppers are included in this grouping when they are harvested for consumption as vegetables and not processed into spices.

Trade data for fresh vegetables also include chilled vegetables, meaning the temperature of the products has been reduced to around 0 °C without the products being frozen. Fruit crops consist of fruits and berries that, with few exceptions, are characterized by their sweet taste. Nearly all are permanent crops, mainly from trees, bushes and shrubs, as well as vines and palms. Fruit crops are consumed directly as food and are processed into dried fruit, fruit juice, canned fruit, frozen fruit, jam, alcoholic beverages, etc.

Source: FAO, Statistics Division
Owner: FAO

GROSS FIXED CAPITAL FORMATION

The gross fixed capital formation is the total value of a producer's acquisitions, less disposals, of fixed assets during the accounting period plus certain additions to the value of non-produced assets (such as subsoil assets or major improvements in the quantity, quality or productivity of land) realized by the productive activity of institutional units.

Source: UNSD, OECD and national statistics' websites
Owner: UNSD, OECD and FAO

IMPORT VALUE

Import values are reported as CIF (cost insurance and freight: the value of the goods, plus the value of the services performed to deliver goods to the border of the exporting country, plus the value of the services performed to deliver the goods from the border of the exporting country to the border of the importing country).

Source: FAO, Statistics Division
Owner: FAO

INDUSTRIAL ROUNDWOOD

All roundwood except wood fuel is industrial roundwood. In production statistics, it is an aggregate comprising sawlogs and veneer logs; pulpwood, round and split; and other industrial roundwood. It is reported in cubic metres solid volume underbark (i.e. excluding bark).

Source: FAO, Forestry Division
Owner: FAO

INFLATION RATE

The inflation rate of an index for any month refers to the percentage change in the index value for the month as compared to the index value of the corresponding month of the previous year. Global and regional food consumer price inflation measures food inflation for a group of countries at different geographical scales: Africa, Europe, Oceania, Latin America and the Caribbean, North America and Asia. Global and regional inflation are calculated using household consumption expenditure weights.

Source: FAO, Statistics Division
Owner: FAO

LAND AREA

Country area excluding area under inland waters and coastal waters.

Source: FAO, Statistics Division
Owner: FAO

LAND AREA EQUIPPED FOR IRRIGATION

Land area equipped with irrigation infrastructure and equipment, in working order, to provide water to crops. The equipment does not have to be used during the reference year. The area equipped for irrigation covers areas equipped for fully controlled irrigation by any of the methods of surface, sprinkler or localized irrigation. It also includes areas under partially controlled irrigation methods of spate irrigation (controlling floodwater to water crops), equipped wetlands and inland valley bottoms and equipped flood recession. It excludes manual watering of plants using buckets, watering cans or other devices.

Source: FAO, Statistics Division
Owner: FAO

LAND UNDER PERMANENT CROPS

Land cultivated with long-term crops that do not have to be replanted for several years (such as cocoa and coffee), land under trees and shrubs producing flowers (such as roses and jasmine), and nurseries (except those for forest trees, which should be classified under "Forestry") are all considered land under permanent crops. Permanent meadows and pastures are excluded from land under permanent crops.

Source: FAO, Statistics Division
Owner: FAO

LAND UNDER PERMANENT MEADOWS AND PASTURES

Land used permanently (five years or more) to grow herbaceous forage crops through cultivation or naturally (wild prairie or grazing land) is considered land under permanent meadows and pastures. Permanent meadows and pastures on which trees and shrubs are grown should be recorded under this heading only if the growing of forage crops is the most important use of the area. Measures may be taken to keep or increase productivity of the land (i.e. use of fertilizers, mowing or systematic grazing by domestic animals.) This class includes:

- grazing in wooded areas (agroforestry areas, for example)
- grazing in shrubby zones (heath, maquis, garigue)
- grassland in the plain or low mountain areas used for grazing: land crossed during transhumance where the animals spend a part of the year (approximately 100 days) without returning to the holding in the evening: mountain and subalpine meadows and similar; and steppes and dry meadows used for pasture.

Source: FAO, Statistics Division

Owner: FAO

LAND USE, LAND-USE CHANGE, AND FORESTRY EMISSIONS

LULUCF covers all GHG emissions and removals produced in the different land use categories, representing the three IPCC Land Use categories: cropland, forest land, and grassland, collectively called emissions/removals from the Forestry and Other Land Use (FOLU) sector. FOLU emissions consist of CO_2 (carbon dioxide), CH_4 (methane) and N_2O (nitrous oxide) associated with land management activities. CO_2 emissions/removals are derived from estimated net carbon stock changes in above- and below-ground biomass pools of forest land, including forest land converted to other land uses. CH_4 and N_2O, and additional CO_2 emissions are estimated for fires and drainage of organic soils. The FAOSTAT emissions database is computed following Tier 1 IPCC 2006 Guidelines for National GHG Inventories.

Source: FAO, Statistics Division

Owner: FAO

LIVESTOCK PRIMARY PRODUCTION

Livestock primary production includes products from live and slaughtered animals. Products from slaughtered animals include meat, offals, raw fats, fresh hides and skins. Products from live animals include milk, eggs, honey, beeswax and fibres of animal origin.

Source: FAO, Statistics Division
Owner: FAO

MEAT

Meat is defined as the flesh of animals (excluding fish) used for food. In production data, meat is normally reported inclusive of bone and exclusive of meat that is unfit for human consumption. As reported by individual countries, meat production data may refer either to commercial production (meat entering marketing channels), inspected production (from animals slaughtered under sanitary inspection), or total production (the total of the above-mentioned categories plus slaughter for personal consumption). All FAO annual production data refer to total production.

Source: FAO, Statistics Division
Owner: FAO

MILK

Whole fresh milk production from buffaloes, camels, cows, goats and sheep.

Source: FAO, Statistics Division
Owner: FAO

NET EMISSIONS/REMOVALS FROM FOREST LAND

Net CO_2 emissions/removals from forest land consist of net carbon stock change in the living biomass pool (aboveground and belowground) associated with: (i) forest, referring to changes occurred on forest land in the reported year; and (ii) net forest conversion from forest land to other land uses. The FAOSTAT data are computed at Tier 3, with the stock difference method, following IPCC 2006 Vol. 4, Ch. 2 and 4.

Source: FAO, Statistics Division
Owner: FAO

NET FOREST CONVERSION, NET EMISSIONS/REMOVALS

Net CO_2 emissions/removals from forest land consist of net carbon stock gain/loss in the living biomass pool (aboveground and belowground biomass) associated with forest and net forest conversion. The FAOSTAT emissions database is computed following Tier 1 IPCC 2006 Guidelines for National GHG Inventories and uses area and carbon stocks data compiled by countries in the FAO Global Forest Resource Assessments.

Source: FAO, Statistics Division
Owner: FAO

NET TRADE

Value in USD of exports minus imports.

Source: FAO, Statistics Division
Owner: FAO

OIL CROPS

Oil-bearing crops or oil crops include both annual (usually called oilseeds) and perennial plants whose seeds, fruits or mesocarp and nuts are valued mainly for the edible or industrial oils that are extracted from them. Oil crops exclude dessert and table nuts, although they are rich in oil, as well as annual oilseed plants that are either harvested green or are used for grazing and for green manure. Some oil crops are also fibre crops in that both the seeds and the fibres are harvested from the same plant (for example coconuts, kapok fruit, seed cotton, linseed and hempseed).

Source: FAO, Statistics Division
Owner: FAO

PAPER AND PAPERBOARD

The paper and paperboard category is an aggregate category. In the production and trade statistics, it represents the sum of graphic papers; sanitary and household papers; packaging materials and other paper and paperboard. It excludes manufactured paper products such as boxes, cartons, books and magazines, etc.

Source: FAO, Forestry Division
Owner: FAO

PESTICIDES

Insecticides, fungicides, herbicides, disinfectants and any substance or mixture of substances intended for preventing, destroying or controlling any pest, including vectors of human or animal disease, unwanted species of plants or animals causing harm during or otherwise interfering with the production, processing, storage, transport or marketing of food, agricultural commodities, wood and wood products or animal feedstuffs, or substances which may be administered to animals for the control of insects, arachnids or other pests in or on their bodies. The term includes substances intended for use as a plant growth regulator, defoliant, desiccant or agent for thinning fruit or preventing the premature fall of fruit, and substances applied to crops either before or after harvest to protect the commodity from deterioration during storage and transport.

Pesticides use data refers to quantities of pesticides applied to crops and seeds in the agricultural sector. Figures are expressed in metric tonnes of active ingredients. However, due to some country reporting practices, the data may be reported by: use in formulated product; sales; distribution; or imports for use in the agricultural sector.

Source: FAO, Statistics Division
Owner: FAO

PREVALENCE OF MODERATE OR SEVERE FOOD INSECURITY

The prevalence of moderate or severe food insecurity is an estimate of the percentage of people in the population who live in households classified as moderately or severely food insecure. The assessment is conducted using data collected with the Food Insecurity Experience Scale (FIES) or a compatible experience-based food security measurement questionnaire (such as the Household Food Security Survey Module – HFSSM). The probability of being food insecure is estimated using the one-parameter logistic Item Response Theory model (the Rasch model) and thresholds for classification are made cross-country comparable by calibrating the metrics obtained in each country against the FIES global reference scale, maintained by FAO. The threshold to classify "moderate or severe" food insecurity corresponds to the severity associated with the item "having to eat less" on the global FIES scale. In simpler terms, a household is classified as moderately or severely food insecure when at least one adult in the household was reported to have been exposed, at times during the year, to low-quality diets and might have been forced to also reduce the quantity of food they would normally eat because of a lack of money or other resources. It is an indicator of a lack of access to food.

Source: FAO, Statistics Division
Owner: FAO

PREVALENCE OF SEVERE FOOD INSECURITY

The prevalence of severe food insecurity is an estimate of the percentage of people in the population who live in households classified as severely food insecure. The assessment is conducted using data collected with the FIES or a compatible experience-based food security measurement questionnaire (such as the HFSSM). The probability of being food insecure is estimated using the one-parameter logistic Item Response Theory model (the Rasch model) and thresholds for classification are made cross-country comparable by calibrating the metrics obtained in each country against the FIES global reference scale, maintained by FAO. The threshold to classify "severe" food insecurity corresponds to the severity associated with the item "having not eaten for an entire day" on the global FIES scale. In simpler terms, a household is classified as severely food insecure when at least one adult in the household was reported to have been exposed, at times during the year, to several of the most severe experiences described in the FIES questions, such as having been forced to reduce the quantity of the food, having skipped meals, having gone hungry, or having to go for a whole day without eating because of a lack of money or other resources. It is an indicator of lack of access to food.

Source: FAO, Statistics Division
Owner: FAO

PREVALENCE OF UNDERNOURISHMENT

Expresses the probability that a randomly selected individual from the population consumes an insufficient quantity of calories to cover their energy requirement for an active and healthy life. The indicator is computed by comparing a probability distribution of habitual daily dietary energy consumption with a threshold level called the minimum dietary energy requirement. Both are based on the notion of an average individual in the reference population.

Source: FAO, Statistics Division
Owner: FAO

PRODUCER PRICES

Producer prices are prices received by farmers for primary crops, live animals and livestock primary products as collected at the point of initial sale (prices paid at the farm-gate).

Source: FAO, Statistics Division
Owner: FAO

PRODUCTION

Figures relate to the total domestic production whether inside or outside the agricultural sector, i.e. they include non-commercial production and production from kitchen gardens. Unless otherwise indicated, production is reported at the farm level for crop and livestock products (i.e. in the case of crops, excluding harvesting losses) and in terms of live weight for fish items (i.e. the actual ex-water weight at time of catch). All data shown relate to total meat production from both commercial and farm slaughter. Data are expressed in terms of dressed carcass weight, excluding offal and slaughter fats. Production of beef and buffalo meat includes veal; mutton and goat meat includes meat from lambs and kids; and pig meat includes bacon and ham in fresh equivalent. Poultry meat includes meat from all domestic birds and refers, wherever possible, to ready-to-cook weight.

Source: FAO, Statistics Division
Owner: FAO

PRODUCTION, CROPS

Production and crops refer to the actual harvested production from the field or orchard and gardens, excluding harvesting and threshing losses and that part of a crop not harvested for any reason. Production, therefore, includes the quantities of the commodity sold in the market (marketed production) and the quantities consumed or used by the producers (auto-consumption). When the production data available refers to a production period falling into two successive calendar years and it is not possible to allocate the relative production to each of them, it is usual to refer production data to that year into which the bulk of the production falls. Crop production data are recorded in tonnes (t). In many countries, crop production data are obtained as a function of the estimated yield and the total area. If such a compilation method of production statistics is enforced by the country, it must be ensured that the total area does not refer to sown or planted areas, which would then give the biological production, but to the actually harvested area during the year.

Source: FAO, Statistics Division
Owner: FAO

RECOVERED PAPER

Waste and scraps of paper or paperboard that have been collected for reuse or trade include paper and paperboard that have been used for their original purposes and residues from paper and paperboard production.

Source: FAO, Forestry Division
Owner: FAO

ROUNDWOOD

All roundwood felled or otherwise harvested and removed is comprised of all wood obtained from removals, i.e. the quantities removed from forests and from trees outside the forest, including wood recovered from natural, felling and logging losses during the period, calendar year or forest year. It includes all wood removed with or without bark, including wood removed in its round form, or split, roughly squared or in other form (e.g. branches, roots, stumps and burls, where these are harvested) and wood that is roughly shaped or pointed. It is an aggregate comprising wood fuel, including wood for charcoal and industrial roundwood (wood in the rough). It is reported in cubic metres solid volume underbark (i.e. excluding bark).

Source: FAO, Forestry Division
Owner: FAO

ROOTS AND TUBERS

Roots and tubers are plants yielding starchy roots, tubers, rhizomes, corms and stems. The denomination "roots and tubers" excludes crops that are cultivated mainly for feed (mangolds, swedes) or for processing into sugar (sugar beets), and those classified as "roots, bulb and tuberous vegetables" (onions, garlic and beets).

Source: FAO, Statistics Division
Owner: FAO

SAWNWOOD

Wood that has been produced from both domestic and imported roundwood, either by sawing lengthways or by a profile-chipping process and that exceeds 6 mm in thickness is sawnwood.

Source: FAO, Forestry Division
Owner: FAO

SHARE OF CEREALS, ROOTS AND TUBERS IN DIETARY ENERGY SUPPLY

The indicator expresses the energy supply (in kcal/cap/day) provided by cereals, roots and tubers as a percentage of the total Dietary Energy Supply (DES) (in kcal/cap/day) calculated from the correponding countries in the FAOSTAT food balance sheets.

Source: FAO, Statistics Division
Owner: FAO

STARCHY ROOTS

Starchy roots include cassava and products, potatoes and products, sweet potatoes and other roots.

Source: FAO, Statistics Division

Owner: FAO

STUNTING, CHILDREN UNDER 5 YEARS OF AGE

Height-for-age less than -2 standard deviations of the World Health Organization (WHO) Child Growth Standards median, among children aged 0–59 months.

Source: World Bank

Owner: UNICEF/WHO/The World Bank: Joint child malnutrition estimates

SUGAR CROPS

Sugar crops include sugar beet, sugar cane, sugar crops nes.

Source: FAO, Statistics Division

Owner: FAO

UNDERNOURISHED, NUMBER OF PEOPLE

The number of people undernourished is obtained by multiplying estimates of the proportion of undernourished for each country by estimates of the total population. Undernourishment refers to the condition of people whose dietary energy consumption is continuously below a minimum dietary energy requirement for maintaining a healthy life and carrying out light physical activity.

Source: FAO, Statistics Division

Owner: FAO

VEGETABLE OILS

Vegetable oils are the oil equivalent of oil crops, which include seeds, nuts, oil palm fruit, olives and soybeans.

Source: FAO, Statistics Division

Owner: FAO

WATER STRESS

Water stress is the ratio between total freshwater withdrawn by all major sectors and total renewable freshwater resources, after taking into account environmental water requirements.

Source: FAO, Land and Water Division (AQUASTAT)

Owner: FAO

WOOD CHARCOAL

Wood carbonized by partial combustion or by heat from external sources is wood charcoal. It includes charcoal used as a fuel or for other uses, e.g. as a reduction agent in metallurgy or as an absorption or filtration medium.

Source: FAO, Forestry Division

Owner: FAO

WOOD FUEL

Roundwood that will be used as fuel for purposes such as cooking, heating or power production is wood fuel. This includes wood harvested from main stems, branches and other parts of trees (where these are harvested for fuel) and wood that will be used for the production of charcoal (e.g. in pit kilns and portable ovens), wood pellets and other agglomerates. It also includes wood chips to be used for fuel that are made directly (i.e. in the forest) from roundwood. It excludes wood charcoal, pellets and other agglomerates. It is reported in cubic metres solid volume underbark (i.e. excluding bark).

Source: FAO, Forestry Division

Owner: FAO

WOOD PELLETS

Wood pellets are made from wood agglomerates produced from co-products (such as cutter shavings, sawdust or chips) of the mechanical wood processing industry, furniture-making industry or other wood transformation activities. They are produced either directly by compression or by the addition of a binder in a proportion not exceeding 3 percent by weight. Such pellets are cylindrical, with a diameter not exceeding 25 mm and a length not exceeding 100 mm. They are assumed to have 8 percent moisture content.

Source: FAO, Forestry Division

Owner: FAO

WOOD PULP

Wood pulp is fibrous material prepared from pulpwood, wood chips, particles or residues by a mechanical and/or chemical process for further manufacture into paper, paperboard, fibreboard or other cellulose products. It is an aggregate comprising mechanical wood pulp, semi-chemical wood pulp, chemical wood pulp and dissolving wood pulp. It is reported in metric tonnes air-dry weight (i.e. with 10 percent moisture content).

Source: FAO, Forestry Division

Owner: FAO

WOOD-BASED PANELS

This product category is an aggregate comprising veneer sheets, plywood, particle board and fibreboard. It is reported in cubic metres solid volume.

Source: FAO, Forestry Division

Owner: FAO

FSC
www.fsc.org
MIX
Papier | Fördert gute Waldnutzung
FSC® C083411

Kolibri 360 GmbH
Ferdinand-Jühlke-Straße 7
99095 Erfurt, Deutschland
produktsicherheit@kolibri360.de